STATISTICS
for
College Students
and Researchers

Michael M Nikoletseas, Ph.D.

Copyright © 2010, Michael M. Nikoletseas
ISBN:9781453604533
Published in USA
Orders:
https://www.createspace.com/3458209

Copyright © 2010 Michael M. Nikoletseas
All Rights Reserved. No part of this book may be reproduced, stored in a retrieval system, or transmitted in any form, or by any means, electronic, mechanical, photocopying, recording or otherwise, without prior permission of the author.

Revised August 2014

CONTENTS

Chapter 1: p. 11
Numbers, quantities measurement
Chapter 2: p. 25
Goddess normal curve
Chapter 3: p. 37
Variance, standard deviation
Chapter 4: p. 48
Uses of the normal distribution
Chapter 5: p. 78
The t-test
Chapter 6: p. 128
Analysis of Variance
    One-Way ANOVA
    Factorial designs
Chapter 7: p. 198
Repeated measures, correlated groups
Paired t-test
ANOVA repeated measures
Complex designs

APPENDICES p. 213

PREFACE

This book is based on the lecture notes and experiences that I accumulated while teaching a two-semester senior course in experimental neuroscience.

The idea, encouragement, and motivation to write the book, comes from my students who have been very successful in entering graduate programs and holding appointments in prestigious schools such as Princeton, Yale, Scripps Institute, MIT, Harvard, Karolinska, Columbia and others. Several of them did their doctoral thesis with Nobel class research teams and now hold faculty appointments.

As for me, I am a poet; I always tried to brush aside feathers and ornaments and look for the heart of the matter.

This book is what remains, if you squeeze a typical statistics book. What remains in the memory of a scientist years after taking Statistics courses. The essentials. The title of this book could have been "Statistics with no formulas" or "Concepts in Statistics", "Statistics for part-time philosophers" or "Statistics for Kindergarten" or any such absurd title.

The most important part of a scientific experiment is the mental and practical activities before the experiment is even started. When the experiment is over, practically all science-doing is over. At this stage, statistical analysis, if needed, will answer a simple question: Is my finding reliable? This simple and obvious fact is often ignored in the frenzy of our times.

You have to love gambling theory, or be compulsive, to like statistics.

Be that as it may.

The trouble is that you and I cannot go against the current. If you plan to go to graduate school and get a masters or Ph.D., you have to know statistics. Knowing statistics means that you understand statistics. Understanding statistics does not mean using a wild array of complex formulas. To the contrary, I believe that formulas are often used in such a way that they blind you and prevent you from understanding the concepts and logic of statistics.

Analyzing your data is easy today.

Most statistics tests are on your computer. They come with it when you buy it. There is also a plethora of web sites that offer you online data analysis. The difficulty is in choosing the appropriate statistical test, and in justifying your analysis, e.g. in presenting your data in meetings, and the oral defense of your thesis, masters or doctorate.

The goal of this book is to teach you the concepts of statistics by walking you through the evolution of statistical ideas in a simple and enjoyable manner.

I will guide you so that you understand the concepts of statistics as they emerged in the early history of this discipline. By understanding I mean that you grasp the concept, the procedure, or mechanism without words, at the gut level, as I say, and, surprise, no formulas! The few formulas (five simple formulas) that you will learn, simply summarize what you already know. You will be able to act out these formulas in you mind, without words. In short, I will teach you with games and stories. It is the game that is important, not the formulas. If you know what you are doing, you do not need a recipe. The formulas, in a real sense, are redundant.

After you read this book and digest the concepts, you can teach fifth graders Analysis of Variance by devising games that they would love to play.

On a more serious level, among the gains of the trip we will take together, will be the changes that will take place in you, and hopefully, you will become a serious and successful scientist.

Oh, I almost forgot. A promise. Every single thing that I will ask you to learn will be necessary for the next step, and will be an important component in the edifice we will build. I will not ask you to learn anything that is not necessary for reaching our goal. All the things that you will learn are part of a pyramid, with Analysis of Variance at the top. You will not be asked to learn something just for the sake of learning it. We will build a pyramid and will use only the necessary materials, no ornaments, no things hanging out of the building.

June 2010

Chapter 1

Numbers, quantities, measurement

How many

Drama
My kids my fingers

A family of early *homo sapiens* sitting around a fire in a cave in Africa, devouring their evening meal of a goat-like animal. The kids are dancing and chasing each other. The women chat and laugh aloud. The man is staring into the dark opening of the cave entrance. Now he raises his hands in front of his eyes and with the index finger of his right hand touches the fingers of his left hand, fixing his eyes on his kids one by one. One kid one finger. He repeats this with the index of his left hand. The next day he is out hunting again. Crouching in the bushes, he waits for the game to pass. Now he brings his hands in front of his eyes and looks at his fingers. He smiles. He knows how many children he has. At the gut level. No words. The dawn of a numbering system is in the air. Thousands of years later, the names of each finger will not be the names of a child, but symbols, words, that indicate the frequency of occurrence of children. Numbers. One, two, three, which are words, mind you. Thousands of years later, these words will be represented by written symbols: 1, 2, 3 and so on. Pretty primitive arithmetic, you will agree.

Drama
No-number numbers

Now lets get transported to a soccer football field in New York. You are watching the game with your friends. The truth is that you are not very fond of soccer so you bring out your laptop computer and play around. Your friends are absorbed in the game, shouting and jumping with excitement. In your boredom you ask: *Who is the guy with the number 11 on his shirt? He must be the best in the team. Poor number 3 he must be one of the worst players.* Your friends are looking at you irritated. Now you add up the numbers on the shirts of the football players. 1 + 2 + 3 + 4 + 5 + 6 + 7 + 8 + 9 + 10 + 11 .
Guys!, you shout, *the sum of the numbers on the football players is 66.*
A while later you use your fancy laptop and calculate the standard deviation. *Guys!* You never finish your sentence, as you friends grab you and throw you out in aisle.

You see that these numbers are strange numbers. They are more like words, *names*. You cannot add, them, subtract them, divide them. They do not express quantity.

These numbers are at the *nominal scale of measurement*. The only advantage of these numbers over names is this: they tell us how many elements we have. They refer to *non-orderable countables*.

Pretty primitive.

Who's bigger

Drama
Who's bigger

Twenty first century Athens. Today the Marathon race for women is taking place. Before you leave home you get a glimpse of the athletes on TV at the start line in Marathon. The end of the race is Athens stadium in downtown Athens. You get back home at dinner time and find your family watching the news. The three winners are proudly standing as the national anthems play. *Great athletes!,* you say. Your family burst into laughter. They are all laughing at you. *Why are you laughing? I said something wrong?,* you say. *This was a great race, especially in the smog of the city of Athens.* They laugh at you even more. Eventually your father explains to you. The first athlete to terminate was indeed great. She broke the world record. However the second to terminate broke the world record in a negative sense. She took so long to terminate the race that no known race has recorded. The smog made all athletes faint.

You see that this type of measurement has an advantage over the previous one (nominal). It tells you what element is biggest, second biggest, and so on. The drawback is that it does not tell you by how much the first differs from the second and so on. You did not know how many hours slower was the second winner in the Marathon race that we considered above.

Numbers in this case are in the *ordinal scale of measurement*. Needless to say, you cannot add, subtract, multiply or divide these numbers. So you cannot figure out means or standard deviation. You cannot run sophisticated statistical analysis.

Primitive.

Equal distance between numbers

Equal distance between numbers

Between 72 and 73 degrees Fahrenheit there are subdivisions, so you can express temperature as 72.6 or 72.4. The distance between 72 and 73 is considered to be the same as the distance between 73 and 74.

You see that this type of measurement has an advantage over the previous two. Most data in the social science are of this type of measures. Here you can add, subtract, multiply, and, practically speaking, divide, However, there is one disadvantage, if you want ot be a purist. Strictly speaking you cannot divide. This will become clear in the next type of measurement.

My zero is not zero

My zero is not zero

In advanced sciences such as physics, measurement is a science in itself. When we say zero in Physics we mean absence of what we measure. And more than that.

Zero in physics must refer to the true state of affairs in the phenomenon we measure. Zero temperature in Physics means absence of molecular movement, since the definition of temperature is molecular movement of a substance. Temperature in Physics is measured with the Kelvin thermometer, which would show zero when there is total absence of molecular movement.

This type of measurement is like the interval type that we considered above, but it is more advanced, since here we can have addition, subtraction, multiplication, and division. This is the *ratio scale of measurement.*

> There are four types of measurement, officially referred to as *scales of measurement*:
> **nominal scale of measurement**
> **ordinal scale of measurement**
> **interval scale of measurement**
> **ratio scale of measurement**

What do you say?

You told us that you will only cover what is practically useful. What use is all of this to me?

Understanding the scales of measurement is very useful. When you are planning your experiment you should spend some time considering what scale of measurement your data will be.

If you can choose interval or ratio scale of measurement, your experiment will be more profitable, and, also, you can run more sophisticated statistics.

When you have collected the data of your experiment, you want to analyze them using statistics. This is a big headache. You must choose the correct statistical test.

Grossly speaking, there are two books of statistical tests: *Parametric*, and *non parametric*.

Here you are, after our first talk, you are in a position to choose one of the books for your data analysis. If your data are nominal or ordinal, you use non parametric statistics, if your data are interval or ratio, you use parametric.

Chapter 2
Goddess Normal Curve

The normal distribution

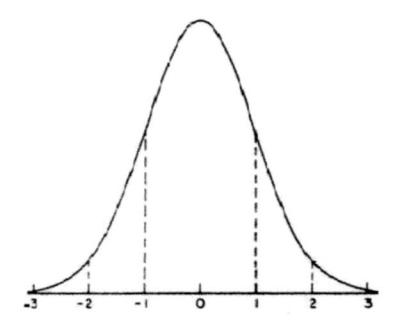

You should bow and pray. This is Goddess Normal Curve. The mother of all. Elegant, but most important, with hidden magic qualities that you can profit from.

In our trips on the barren land of Statistics, in times of despair, we will ask her for help and inspiration. Let me say this in other words. All the reasoning we will engage in in this book, will be while we stare at this goddess, as we scratch our heads in search of a solution.

Do not get discouraged by looking at the graph above. Remember fifth graders can understand this.

Ok, let's take a good look at it. Notice that it looks like a Texas hat. It is symmetrical, meaning that, if you take a pair of scissors and cut it in the middle, the two parts are identical, or better, mirror images.

It is perfect, isn't it?

Like a true goddess, the normal curve does not exist in our material world. What we see is simply an imperfect reproduction of it. The real normal curve exists in our minds, it is a concept. Mathematicians have produced a formula which makes this graph (see Appendix 3).

Now watch that in the middle of the horizontal axis there is a 0. On the right side of the midline 0 there are 2 vertical lines. There are also 2 vertical lines on the left of the midline 0. (I know you know this without looking, since the two parts are identical).

Drama
Working magic with a Goddess

Now get your fifth grade pupil, Tom, and run an experiment. Draw this curve on a card board. Take a pair of scissors and cut this curve out so that you really have a piece of cardboard that looks like a hat. Make sure this hat weighs 100 grams.
Now ask Tom to weigh the cardboard hat.

It is 100 grams, he says.

Ask Tom to cut the cardboard in the middle. Ask Tom to weigh the right side. He will refuse, he will say:

It is 50 grams, giggling.

Ask him to weigh the left side. He will laugh again. Now ask Tom to cut along the first vertical line on the *right* of the midline cut. He does and holds the strip of cardboard in his hands. Ask him to guess its weight. His pride is deflated, he does not know. You do not know either. Weigh this strip, You will find that it weighs 34 grams.

Ask Tom to cut along the first vertical line on the *left* half of the hct, and guess its weight. He will refuse.

It is 34 grams, he will say.

Next give Tom a new cardboard cutout of the normal curve, and ask him to cut along the second vertical line on the right side of the midline, and also cut along the second vertical line on the left of the midline on the left side. He holds three pieces now, the big middle part, and the two, tiny pointed parts, the "tails" of the curve. Ask Tom to weight the big middle part. He does.

It is 95 grams, he says.

Now ask him to weigh the two small pieces of the cardbocrd, the "tails". He will refuse, and laughingly he will say

5 grams.

Ask him to weigh one of the small pieces, or tails. He will refuse.

2 and a half grams, he will say.

Child's play, you agree?

If you were to ask a mathematician to do this experiment for you, he would laugh at you right away. Mathematicians can calculate this from the formula that generates the normal curve.

The numbered points of the vertical lines, 3 on each side of the midline, they call "standard deviations". Standard deviation +1, standard deviation +2, standard deviation +3 on the right side, and standard deviation -1, standard deviation -2, standard deviation -3 on the left side.

We do not have to worry about this. What we should always remember is that that if we cut the cardboard model of the normal distribution along the second vertical line to the right and left of the midline (that is at standard deviations +2 and -2), the bigger piece weighs approximately 95 grams, and the two small pieces, the "tails" weigh approximately 5 grams.

If we cut along the third vertical line from the midline (standard deviation +3 and -3, the middle piece weighs approximately 99 grams, while the two outward noses or tails of the curve weigh approximately 1 gram, each 0.5 grams.

In other words, approximately 95% of the cardboard weight is between standard deviation +2 and -2, and 5% of the cardboard weight is beyond standard deviation +2 and -2.

The weight of the cardboard between standard deviation +3 and -3 is approximately 99% of the total, and, obviously, 1% of the total weight is the cardboard piece beyond +3 and -3. These are approximate values.

It is important that you remember this.

Approximately, you say. *What are the exact values?*

Standard deviation 1.96
Percent of curve 95%
Standard deviation 2.58

It is important that you remember this.

Let me ask you something. What is the surface area of the cardboard, between standard deviation -2 and +2? Amazingly, most students cannot answer this question. I know that some of you see this as a silly question.

Of course, if the weight of the cardboard between -2 and +2 standard deviations is approximately 95%, the surface is also approximately 95% .

Now we will play a variation of the same game. We will imagine that the cardboard model of the normal distribution is a Texas hat.

We stand at a corner in downtown Houston and ask a man passing by to, please, allow us to measure his height. We write his height on a small piece of paper, and throw it in the hat, and of course we also type it in a laptop computer. After we measure the heights of 1000 men, we get into the boring task of arranging the small pieces of paper in the hat, stacking them up. The smallest numbers go to the extreme left, and the large numbers to the extreme right, and the rest, the middle of the road values, in between. As you would expect, there will be very few men that are very short. Most men will be around a middle value, the average, and again, very few men will be extremely tall. Look at the normal distribution. As we move from left to right the height of the curve increases, it reaches a peak exactly in the midline, and decreases again on the right side. You see that the height of the curve indicates *frequency* (how many times a given height occurs).

Below the tallest point of the curve is the mean, the average. This, as you can easily see is also the most frequent score, the *mode*, that is the most fashionable score.

The scores (the small pieces of paper with the heights of Texans) between standard deviations +2 and -2 represent approximately 95% of the total number of scores we put into the hat. The scores between standard deviation +3 and -3 represent 99% of the scores. Again these are approximate values.

Let us pause for a moment and see where we are and what we are doing.

We said that the normal curve is a kind of goddess that helps us reason in statistical problems. We saw some of its characteristics and saw that the markings of standard deviation help us calculate the percentage of the weight, the surface area, or the frequency of scores of this curve.

What do you say?

You told us that we will understand perfectly every step and every concept as we go along, you say. *I do not understand what standard deviation is.*

Ok. We will develop the concept of standard deviation right away.

Chapter 3
Variance and standard deviation

Drama
Mathematical sweat

Susan Bolles is a new professor of Psychology at Goatshead College. Her chairman, sorry, chairperson, Dr. Alexa Terrorvski, assigned her the introductory psychology class of 1000 students. The first midterm exam has just taken place. There were 100 questions, 1 point each. The exam papers were computer-graded, all 1000 of them. Dr. Terrorvski wants to know how the class did, so she asks Susan. Susan says that the mean (the average) was 60.

Dr. Terrorvski wants to know more, how many people scored close to 100, and how many people scored close to zero. Susan walks up to the pile of exam sheets and starts reading the scores: *48, 30, 70, 99, 53* ...... Papers are spread out from the middle point, the average.

Realizing that this would take a good part of the day, Dr. Terrorvski shouts out:

*There must be a better way!
I will tell you what. Take this pile out to the stadium, put it down in the middle of the stadium. Mark this point 60 (your average score). Take a step and mark this point 61. Another step, mark this 62, all the way to 100. Return to mark 60 and take a step in the opposite direction. Mark this point 59. Take another step, and mark this 58, repeat all the way to mark 0. Now return to your pile of exam sheets and pick up an exam sheet. Read the score and walk to the point it corresponds to on the markings you made. I will return in an hour to see how the exam went.*

When Dr. Terrorvski returns she finds Susan drenched in sweat and panting vigorously. There is a long line of white sheets of paper on both sides of the point that marks 60, the average. Susan picks up another exam sheet which had the score of 3 and begins walking. 59. 58. 57

Enough!, Terrorvski shouts. *This is a mess. Look at how many papers are spread out away from the mean, students have scored low scores, all the way down to zero. So many scores, so many students are very far from 60, the mean. There is a big distance of many scores from the mean. You need to be more effective in teaching your students, even the weak ones,* Dr. Terrorvski said, and marched out of the stadium.

Susan went to her office and tried to get a better picture of the situation. Rather than walking away from the point of the mean, she calculated the distance of each score from the mean. Score 40. Distance from the mean -20. Score 65. Distance from the mean +5. And so on. At the end she added up all of these distances and she found the total distance. That was the distance she had to walk in the stadium!

In the second midterm, the mean was again 60. This time she did not go out in the stadium. She simply found out the distance of each score from the mean. She added up all of these distances and was pleased to see that the total distance was very small. Surely, Dr. Terrorvski would not yell at her this time. Students scored close to the mean, there were very few low scores. The scores this time were not spread out all over the place away from the mean.

Let's see what Susan did. She calculated the average, the mean. As even fifth grader Tom knows, we find the mean by adding up all scores, and divide by how many scores went into the calculation.

The formula for this is

$$\bar{X} \approx \frac{\Sigma X}{n}$$

We read this as follows: X bar, equals the sum of X divided by n.

X bar is the symbol for the mean.
Σ is the symbol for addition, sum.
X is the symbol for score. n is the symbol for how many scores.

After calculating the mean, Susan. calculated the distance (the deviation) of each score from the mean. In doing this she calculated the total distance of all scores from the mean. That is she calculated a measure of the spread, or dispersion of scores around the mean. That is a measure of variation or variance of scores.

The symbol for variance is *s* 2

The formula for variance is

$$s^2 \approx \frac{\Sigma(X-\overline{X})^2}{n}$$

We read this as follows: Variance equals the sum of squared deviations of each score from the mean, divided by how many scores went into the calculation.

Why squared? Why square the deviation, you say.

The sum of deviations from the mean *always*, in all cases, equals 0. That is why we square each deviation to prevent this. You should know that in all sciences, for the purpose of meaningful analysis, we may transform our data by squaring them, or expressing them as logarithms, and so on. This does not change the relation of scores amongst themselves.

Why divide by n?

You understand that if in one case we have large scores, and in another small scores, the sum of the deviations from the mean will be large in the first instance, and small in the second instance. If we want to compare the spread of the scores in the two instances, we have to average each of these sums of deviations. I hope you understand this.

For example, in order to compare the income of New Yorkers to that of Chicagoans, we must average the total income of New Yorkers and also Chicagoans. The numerator of this formula

$$\Sigma(\overline{X} - X)^2$$

is the sum of the difference of each score squared, or raised to the second power. More formally, we say the the numerator of the variance formula is the sum of squared deviations of each score from the mean, squared. In statistical jargon we say: Sum of squares, or SS.

> The numerator of the variance formula is the Sum of Squares, or SS

The denominator is the n, i.e. the number of scores we have in this case. Dividing by how many scores we have the mean or average.

So, the variance formula is the average of the sum of squared deviations, or, in statistical jargon, the mean squares or MS, for short.

> Variance is also called *mean squares* or *MS*

What is standard deviation? you say.

To calculate the standard deviation we take the square root of variance. Simple.

You do not need to know how to calculate the square root of a number. Not in the age of computers. Anyone can learn to do simple arithmetic. The challenge is to understand concepts of statistical and mathematical operations.

The formula for standard deviation is:

$$s \approx \sqrt{\frac{\Sigma(X-\bar{X})^2}{n}}$$

Do not worry about the -1 in the denominator. The n changes depending on whether we deal with samples or an entire population. Remember our goal here is to understand the concepts of statistics and want to avoid getting stuck in compulsive swamps.

Now that we have removed the mystery of standard deviation of the normal distribution, we return to it.

Remember this is not just a curve, it is Goddess Normal Curve. Glory to NC in the highest!

Chapter 4
The uses of the normal distribution

Faithful to our plan not to get sidetracked and talk about just about everything statistics books talk about, we put our foot down and say:

Why learn all of these things about the normal curve? What is the use of all of this?

The use of all of this is necessary, I say.

The normal curve is a mathematical, perfect curve, with magic qualities and powers. Understanding the normal curve is necessary, if we wish to understand statistics from simple t-tests to complex Analysis of Variance (ANOVA).

The normal curve is used in four instances:

1. To describe, to organize data.

2. To make statements regarding probabilities as to the occurrence of a particular score, as in games of chance.

3. To make statements regarding the reliability of a single mean

4. To make statements regarding the reliability of the difference between two means.

Understanding the concepts in number 4 above is the basis for understanding the concepts of all statistical tests. Also, as we said before, there is a continuity in the process of our understanding of statistics.

It is like a fairy tale. You must know the full story, starting from the beginning and step by step reach the end, in order to make sense. So keep alert!

Normal distribution, use number 1.
To describe, to organize data

Michelangelo, Sistine Chapel

Point of contact.
God's hand makes contact with the hand of Man. Genesis. Magic moment. A whole world begins here. The divine, the immaterial, the perfect makes contact with the earthly, imperfect, and imparts to it some of the harmony of the spiritual, perfect world.

Drama
Where does Basita fall?

Susan, our psychology professor, decided to take a personal interest in the learning of her students, and called those scoring very low to her office. Among those she called was Basita.

The bottom line of this is that you should quit college immediately. You are the bottom of the bottommost. You will never be able to compete with other college students. Find a job in a diner, in a farm, anywhere, but do not waste your time at college, she said to Basita.

The next day, Basita and her mother, Mrs. Thinlips, an accountant by profession, marched into Susan's office.

I have already talked to your chairperson about this. I demand that you explain to me the basis of your criticism and absurd advice to my daughter. You traumatized her, in effect telling her that she is an idiot. You will hear from my lawyer. For now I want an explanation.

My daughter scored 45. The mean was 60. Forty five is close to the mean, only 15 point below. Forty five means that Basita knows almost half of what you expect her to know. Your telling my daughter to quit college is most unwarranted. I demand an explanation!, Mrs Thinlips said, banging her fist on the Susan's desk.

Help, Susan said to herself, *Goddess Normal Curve, help.* She brings out a sizable cardboard model of the goddess and bows.

Mrs, Thinlips, she said. *The mean of the scores in Basita's class was indeed 60, and the standard deviation was 5. Here is the computer analysis.*

Now we place 60 on the mean (0 standard deviation), that is in the middle of the curve.

Flash, thunder, tempest winds, Michelangelo hovers over the cardboard model! Angels and ministers of heaven and hell! Point of contact of the spiritual with the material! A new science is born. Statistics. All else is humble things after the cosmogony of the this moment of Genesis. All subsequent statistical tests bow to this archetypal creation.

We place 60 on the mean, that is in the middle of the curve, Susan continues. *Now we move down to standard deviation -1, to the first vertical line on the left of the midline. This means that at this point we have score 55. Now we move down one more standard deviation, standard deviation -2. Here we have score 50. Finally we move down one more standard deviation, standard deviation -3. Here we have score 45. This is Basita's score. The percentage of scores above this point is 99.5%. That is one student out of 200 scored 45 or lower. Since we have 1000 students in this class, no more than 5 students scored the same or lower than Basita. Imagine a line of 1000 students, a small town, and your daughter standing at the very end!* Susan said, with a malicious

smile on her face.
Mrs. Thinlips or Basita have not been seen on the campus ever since.
Back to our task to understand the normal distribution, to understand it our way, a gut-level understanding.

In doing science we have two domains, two worlds. The *empirical domain*, the mud and flesh domain, and the *formal domain*, the domain of abstractions, ideas, logic and mathematics. The empirical domain is our sense world, and the data we get by running experiments in it.

The formal domain is the world of thought and mathematics. Sciences progress by superimposing perfect models of mathematics on the imperfect, variable, messy world of matter. When we do that, we immediately see things that we could not see by looking only at the data we have collected from observations in the material world.

Newton succeeded in creating a revolution in Physics by first creating a calculus, which he superimposed on nature. Galileo Galilei, the man who started science as we know it today, said that the language of nature is mathematics.

A most important note in Basita's story:

What if Basita's score was not 45 but it was 43? How would we find where it falls on the normal curve? There is a formula called the z formula. Here it is:

$$z \approx \frac{X - \overline{X}}{s}$$

Let's try it.

Score 43 minus the mean, which is 60, equals -17. Now if we divide -17 by the standard deviation which is 5 here, we get a z of -3.4. That makes sense. Basita's score of 45 fell exactly on standard deviation -3, as we saw. A score of 43 will be even more to the left of the curve.

I do not want to close this talk. I want to play some triumphant march, Beethoven's Eroica perhaps. Look at this formula. Play with it, do things with it. Digest what we do with it. Let's dramatize this.

$$z \approx \frac{X - \overline{X}}{s}$$

Drama
An archetypal ceremony

I pick a score, and wave it in the air. Then I wear my glasses and stick my nose on the normal curve, running up and down the line with standard deviations on it, I mumble:

where does this score fall? Where does this score fall?

I then use the z formula and find where exactly my score falls.

This is an archetypal ceremony. Remember it. We will act it out again in the future.

Normal distribution, use number 2.
Making statements of probability, betting

Drama
Money in a Texas hat

After taking my course, Nick, an entrepreneurial mind, decided to go into business. He went south, to Houston Texas, and planned a betting business without any substantial investment. Just an antique Texas Instruments calculator. For two months he stood at a corner in downtown Houston asking every man that appeared around the corner:

Excuse me, sir, would you mind if I measure how tall you are? I am running my thesis and need data.

He carefully recorded the data. At the end of the two months he had measured the heights of 4000 men. Now he punched his data into the calculator and computed the mean and the standard deviation.

The mean was 170 cm, that is 1 meter and 70 centimeters. The standard deviation was 10.

The next morning he puts on his brightest face, and stands at the same corner in downtown Houston. Time to make money.

Excuse me sir, I bet $1000.00 that the first man that will appear around the corner will be between 1 meter 50 centimeters, and 1 meter 90 centimeters tall.

Not all passersby pay attention to him but a few do.

Oh Yeah? How do you know, buddy? You think you are smart, ah? Here is $1000. Show me yours.

Tom puts down his $1000. Here he comes, first man appears around the corner. He agrees to be measured. His height is 170. Nick wins. Nick will make several thousand dollars on his first day. He loses a few times but 95% of the times he wins.

Let's see his reasoning.

He followed my example of Basita's story. He placed the mean of the heights,170 cm, (the one that he computed from his data) on the middle of the normal curve. Now he reasoned that since the standard deviation of his data was 10, at standard deviation -1 score 160 exists, at standard deviation -2 score 150 exists, and at standard deviation -3 score 140 exists.

Similarly, on the right side of the curve, score 180 is at standard deviation +1, score 190 is at standard deviation +2, and score 200 is at standard deviation +3.

One more example:

Mean 20
Standard deviation 3

What standard deviation score 26 lies at?

Answer: Score 26 lies at standard deviation +2

At standard deviation +1 we have score 23, i.e. 20+3

At standard deviation +2 we have score 26, i.e. 20+3+3

At standard deviation +3 we have score 29, i.e. 20+3+3+3

At standard deviation -1 we have score 17, i.e. 20-3

At standard deviation -2 we have score 14, i.e. 20-3-3

At standard deviation -3 we have score 11, i.e. 20-3-3-3

Normal distribution, use number 3.
To make statements regarding the reliability of
a
single mean

Drama
A Cap for Wisconsin Farmers

Mike got his degree in Psychology from UW Madison. Given the large number of psychology graduates and also his doubts regarding his suitability for psychology practice, he found a job as a consultant with the State of Wisconsin. Psychologists, you should know, learn a lot of statistics. Around the middle of last November his boss walked into his office and said:

Mike, I have a job for you. The governor has decided to give a present to all farmers in the State of Wisconsin because they are very angry with the new taxes. The gift will be a woolen cap. We want to know the size of the cap. If we can find the average head size of Wisconsin farmers, we can give the order to a factory in Milwaukee to make the caps. They are woolen so they naturally stretch. If we know the average head size we will be ok. Since we only have one month to complete this project, I expect you to report to me with a plan and budget tomorrow.

Early the next morning Mike walked into his boss's office and handed him the proposal. Three hundred personnel to cover the entire state, to locate *every* farmer, in one weak. Fifty 4x4 Jeeps to safely travel to even the remotest towns. One small aircraft to land in the northern towns in case of snow. Three hundred laptops. Ten German Shepherds to smell the bears up and around Wausau. Budget. $30,000.00.

His boss looked at Mike for 20 seconds speechless. Then, in a completely unemotional voice, he said:

Mike, in the State of Wisconsin we are very careful with our money. No way. Find a less expensive way by tomorrow.

Mike began to fear for his job. All day in his office, all night in his home, he scratched his head, drank a lot of coffee, and prayed to Goddess Normal Curve.

*I am not allowed to measure **all** farmer heads. That is too expensive. I can, perhaps, still find a way to use the normal curve. If I can come up with a mean that has a strong probability to be close to the real mean... If I take measures of the heads of many farmers and compute the mean.... Can I be sure that this is close to real mean? If not I will lose my job. So, what if I go out a second time and repeat the data collection, just to make sure that this mean was not a mean that I got by chance but it was a mean close to the real mean. Ah, that might be it! I go out several times and each time I compute the mean. In the end I graph these means (as though they were scores).*

Then I use the normal curve to reason in some way. How? Let me see... The normal curve would be graphing means. So the mean would be the mean of the means. Can I play the game of Nick? He was making probability statements, predictions, regarding the occurrence of scores, using the standard deviation of the scores. Ah! I can use the standard deviation of the means. Then I can reason, like Nick, that a mean close to the mean of means, that is between standard deviation -2 and +2 would have a high probability of occurrence. That's it!.

He prepares the budget, and early in the morning he busts into his boss's office.

It will cost us $10,000, he says.
Good idea, but too expensive. Tomorrow is the last day, boss says, and with the palm of his hand he points at the door.

Three in the morning, Mike is on his knees in front of Goddess Normal Curve, buried in statistics books and statistics journals, and notes from his stat class. Suddenly he comes across an article in a journal which claims that you can calculate an estimate of the standard deviation of the normal curve that would be graphing means.
... *that would be graphing means.. Would be...,* he repeats this several times.

Would be, because this curve has only one mean. Let me say it in another way. You go out and you collect data from a large sample. You can calculate an estimate of the standard deviation of the curve that would be graphing the means of samples that you would be getting, if you were allowed to collect several samples.

Weird..., Mike mumbles. *What good is it? I want to be able to collect one sample, calculate the mean, and tell my boss that we can trust this mean as being close to the real head size of Wisconsin farmers, that it is reliable. What good is computing an estimate of the standard deviation of a curve and not know much else about this curve...*

The traffic noise picks up, it is six o'clock in the morning. Another look at the Goddess, and a supplication for inspiration.

All I know is what Nick did, Mike says. *He placed the mean of his data on the normal curve mean (middle). Unfortunately I do not have the mean of the means, since I am allowed to take only one mean. Let me place the letters TM in the middle of the normal curve, TM for True Mean. TM will remain forever unknown. Pretty spooky. But I can place the standard deviation of this "I-would-be-getting-curve", an estimate of the standard deviation, to be exact.*

Ok, then what.

Weird things happen to people under stress and in despair. Some people hear voices, others are visited by angels, others write poetry…

Got it!, he suddenly exclaims, raises the normal curve over his head, and dances a cannibal dance around his desk.

Eight in the morning he rushes into his bosses office.

One day, one sample, one mean, one thousand dollars!, he yelps.

His boss pretends he is not listening.

I will go out, one day, collect many head size scores, calculate this mean. Next I will compute an estimate of the standard deviation of this curve that you did not allow me to get the data for. I will then run down two standard deviations from the middle of the curve (-2 to +2).

Mike pauses to get some feedback from his boss. Stone silence.

Grant me this, Mike continues in a loud voice. *This curve would be graphing means, right? My one mean is one of these means, right?*

That is absolutely correct, and also tautologous, boss says, and looks at Mike with contempt.

What is the chance that this mean would be one of the 95 percent of the means?, Mike asks.

It highly probable, almost certain, boss replies.

Then the problem boils down to the size of the standard deviation of this curve, i.e. the estimate that we will compute. If the standard deviation is large, then we would run the risk of producing caps that are ridiculously large or small for the heads of Wisconsin farmers. If the standard deviation is small our mean would almost certainly be close to the true mean, and we are in business.

Mike carried out this project successfully without any problems, except that he was chased by a playful bear at *Wausau* up north.

The formula for the calculation of the estimate of the standard deviation of the curve that we would be getting, if we were allowed to get many samples, but are allowed only to take one sample, and so have only one mean, is:

$$SEM \approx \frac{s}{\sqrt{n}}$$

We read this as follows: standard error of the mean equals the standard deviation (of the data from our one sample), divided by the square root of the number of data that go into the calculation, i.e. the n. Yes, you guess right, the official name for the estimate of the standard deviation of the curve that would be graphing means is called *standard error of the mean*.

Normal distribution, use number 4.
To make statements regarding the reliability of the difference between two means

A psychologist at the University of California published a study in which she claimed that college students who prefer Polish sausage react faster as compared to college students who eat plain hotdogs. She measured the time it takes to respond when a stimulus, a buzzer, is presented.
Here is a summary of the data:

Reaction Time ms

	Polish Sausage	Hotdog
Mean	210	215
Standard	20	25
n	200	200

The difference between the two means is 5 milliseconds. The Polish sausage group responds in less time, that is this group is faster. However, a question pops up: Is this difference reliable?

Which means, will we find this difference, if we run the experiment again, or this difference was perhaps found by chance.

We can use the normal distribution to solve this problem. However, it is the almost universal practice in many sciences to use a the t-test, the so called Student's t-test. The t-test was created by William Sealy Gosset.

Chapter 5

The t-test

Wikipedia Aug, 2014 PD
William Sealy Gosset
(June 13, 1876–October 16, 1937)

William Sealy Gosset published the t-test under the pen name Student. We refer to the distribution of this test as the t-distribution. The t-test is *a test of inference*, i.e. it allows us to infer on the basis of our data, whether the difference between two means is reliable or, as we say, *significant*.

In what follows, we will first try to develop the concepts needed for understanding the logic and the operations involved in the t-test.

We will talk about this and that and the other. Be patient. Then we will go over an example of the t-test.

Developing the concepts in the t-test

As is the case with all parametric tests that we will cover in this book, the t-test analysis is based on variance.

In experiments in which we have two groups we analyze our data by using the t-test. There are two types of t-tests. The t-test for independent samples (groups), and the other for dependent or paired samples. Here we will consider independent samples.

What are independent samples? you say.

Ok. We will make a small parenthesis in order to develop the concept of independence.

Drama
Apple-pie IQ

A psychologist has a sneaky suspicion that the type of apple pie has an effect on intelligence. She randomly selected 20 students and randomly assigned them to two groups. Group 1, golden delicious apple pie, Group 2, red delicious apple pie. John Gluck was assigned in Group 1, and his friend Paul Crust was assigned to Group 2. The psychologist proceeded with giving these subjects a pound of apple pie to eat. Subsequently she tested their intelligence. Each subject was allowed to see their intelligence score. There were 20 intelligence scores, one for each subject. These groups are independent as you see. She analyzed her data by using a t-test for independent groups in order to see if there was a significant difference in intelligence in the two groups.

Note: An unexpected event occurred during the running of the experiment. One of the subjects in Group 2 did not show up on time so the experiment was delayed for a few minutes. John Gluck, who, as you remember was in Group 1, offered to participate in Group 2, in addition to his participation in Group 1. The experimenter did not allow this. Had she allowed John Gluck to be a subject in both groups, she would have violated the rule of independence, and she would not have been able to analyze her data by using the t-test for independent samples.

The formula for the t-test is:

$$t \approx \frac{\overline{X}_1 - \overline{X}_2}{\sqrt{\frac{s_1}{n_1} + \frac{s_2}{n_2}}}$$

We read this as follows:
T equals mean 1 minus mean 2 divided by the square root of the variance of group 1 and group 2 divided by the number of scores that went into the calculation of the variance.

Faithful to our goal we must understand the concepts in the t-test.

First look at the numerator. Mean 1 minus mean 2, that is the difference of the two means. Next look at the denominator. The square root of the variance is the standard deviation. This looks like the z formula that we considered above. Here it is again:

$$z \approx \frac{X - \overline{X}}{s}$$

Yes, you say, but the numerator of the z formula is score minus the mean. The numerator of the t-test formula is mean 1 minus mean 2. Where is the mean? They are not the same, you say.

They are the same, I say. Mean 1 minus mean 2 is the difference between the two means. Gosset treats this difference as a score.

Yes, you say, but then where is the mean in the t-formula?

The mean is there, I say.

It is there but you do not see it. It is 0. The mean is zero.

The mean is zero!

Let the drums thunder at this point. Let the bugles sound in the four corners of the world!

The normal (t-) distribution with 0 in the middle. In other words a curve with a mean of 0. A most important point in the history of statistics. Let us call this curve the *curve of no difference*.

I do not understand the t formula at the gut level, you say.

Watch the ritual dance with the t-formula.

Drama
An archetypal ceremony II

I have a difference between the two means. I hold this difference up, wave it in the air, *I baptize it score*. Then I wear my glasses and stick my nose on the t curve, running up and down the line with standard deviations on it, and mumble: *where does this score fall? Where does this score fall?*

I then use the z formula - oops, the t-formula - and find where exactly our score - oops, our difference - falls.

This is an archetypal ceremony.
Remember? Amazing, isn't it? The
t-formula is actually the z formula!
I promised you that you do not
need the mind boggling array of
fear inspiring formulas. Hang on.

Here is more of the story of the t
distribution. The motive in
Gossets' mind was to modify the
normal curve so that it could safely
be used for small samples. He
decided to make it difficult for
researchers to find significance
(i.e. to decide whether the
difference between our two means
is reliable) when the samples are
small. The curve he created is a
normal curve with some intriguing
qualities. The noses, or tails of the
curve lift up as the sample size
decreases. The tails of the curve
lift up.

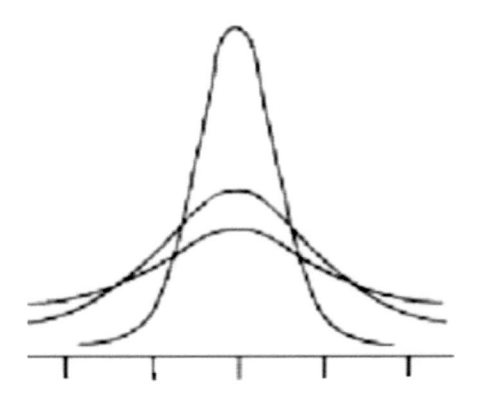

The shortest curve corresponds to the smallest sample. What an ingenious idea! Some of you, few, very few, already see what this means. It means that it becomes more difficult for you to find significance because the percentage of the curve increases in the tails so that the magic standard deviation of 1.96 becomes larger, which in turn makes it difficult for you to find significance, that is to have a "real" effect, a "true" finding so that you can publish your experiment.

Let's finish this story of the t-test,

a statistical test that is used so frequently in labs across the world daily.
So far we have said that we have two groups, therefore two means, two standard deviations. We are interested in deciding whether the difference between the two means is reliable, i.e. that it is not a chance event, that it is, in a sense, real. We find the t, which is actually a z, that is, it tells us where on the curve this difference falls.

Ok, you say. *So we find the t which is like z, then what?*

If our sample is large the t distribution is identical to the normal distribution. In the normal distribution, we have seen that z of 1.96 is an important mark. Between -1.96 and +1.96 95% of the curve falls. A score (here a difference) that falls within these two marking points, has a probability of 95% to occur by chance in a known situation in which there is no difference, that is in a situation in which the two samples were drawn from the same population. That is what our curve of no difference graphs.

If our sample is small? you ask.

While in the normal distribution 1.96 always marks the 95% of the curve, in the t distribution it is so *only* if the sample size is very large. With smaller samples, 1.96 *increases* inversely proportional to sample size. The smaller the sample size, the greater the increase in 1.96.

Remember, again, Gosset's goal was to make it difficult for researchers to find significance with small samples. He calculated these new values of 1.96.

Drama
Mercy Mr. Gosset

Mr. Gosset, good morning. This is Samir calling from India. I ran an experiment with 20 subjects. How much should I increase 1.96?

Half asleep, Gosset takes out his notes and read the value to Samir.

The new value for z 1.96, the t, is 2.101.
Good night Mr. Samir.

Two hours later the phone rings again.

Good evening, Mr. Gosset. I am Michel Duzed, calling from Montreal Canada. Please give me the value of z for an experiment with 72 subjects. Gosset looks at his notes and says:

It is 1.994
Goodnight.

Going back to sleep is difficult for Mr. Gosset. One hour later the phone rings again, this time from Japan.

Good day Mr. Gosset. Please give me the new z for an experiment with 402 subjects.

1.966 , Gosset says and he hangs up. *I got to do something about this,* he says, *aside from pulling the phone from the plug. I got to do something... I will never be able to sleep.*

He did. He published his notes
with the recalculated z of 1.96 .
Have you seen this in any
statistics book? All of my students
say no, actually a lot of my
colleagues say no, too. I will
disclose the secret. It is the
famous t-table found at the end of
every statistics book, including the
one you are reading now (see
Appendix 1). Promise to keep our secret
between us.

Climax in the drama.

Samir of India wrote down the t value: 2.101

With hands trembling he picked up the sheet with the data analysis of his thesis to find the result of the t-test. He read it out aloud:

t=2.24
I made it! I made it!,

he chants as he dances around in his room. Samir will get his Masters. The difference between the two means of his experiment is reliable.

He will report his finding as significant. In his thesis he will write:

This difference is significant ($p<0.05$).

What does this notation, (p<0.05), mean?

It means that the chance that his finding is not reliable, i.e. that it is a chance event, is less than 5 per cent.

Scientists have agreed to accept findings as being reliable, if the p value is less than 0.05.

> Scientists have agreed to accept findings as being reliable,
> if the p value is less than 0.05.

Remember this!

Back in Canada Michel Duzed, holding the note with the t value that Mr. Gosset just gave her (it was 1.994, remember?), compares it with the result of the t-test value that she got by analyzing the data of her experiment, which was t=1.982.

Alas! The t she computed by using the t formula is smaller than the t Mr. Gosset gave her.

Her eyes open wide, her face get gloomy, and she collapses on an armchair. Michel will not get her Doctorate. Her finding is not significant. She will not write her dissertation. If she were to write it, she would report it as follows.

This finding is not significant (p>0.05).

This means that her difference could have been a chance event more than 5 times in a hundred.

Michel quits graduate school, she marries her sweetheart and moves out of Quebec to a job as a business consultant.

What happened to the Japanese guy?

His finding was significant too. But he never got his Doctorate because he refused to write up his thesis. He joined the ABD club. Eventually I hear he became a famous poet. What does ABD mean? Figure it out, or wait till you are about to write up your doctoral thesis yourself.

Encore the concepts.

Before giving an example of the t-test, we will launch a final attack on the conceptual difficulties of the t-test and inferential statistics in general.

First let us refresh our memory.

Between standard deviation or z -1.96 and +1.96, 95% of the curve lies. It follows that the remaining of the curve beyond -1.96 and +1.96 represents 5% or the curve.

> Standard deviation 1.96
> Percent of curve 95%
> Standard deviation 2.58

The curve of no difference, as we said, has zero in the middle, that is the mean is zero. What does this curve graph? Remember the curve in the case of the woolen caps for the farmers of the State of Wisconsin. That curve was a curve that we would be getting if we were allowed to collect many samples, figure out the mean of each, and graph these means. In effect this curve was empty.

But, remember we knew something about it. We knew the estimate of the standard deviation (standard error of the mean). We calculated it from the standard deviation of the one and only sample mean we had.

In our present case, the t-curve is a similar curve. It would be graphing differences between two means, if were we allowed to run our experiment of two groups many times, each time having two means, calculating the mean of each group and then calculating the difference of the two means.

Because we have the standard deviation of this curve, we can consider our difference as a score and engage in the ritual dance of where my score falls. What point on the standard deviation line does this score (difference) lies. We used the z formula to tell us where on the curve our difference lies. We said above that the t-formula is actually a z formula, a modified one to take into account the sample size. Let's look at them again.

$$z \approx \frac{X - \overline{X}}{s}$$

$$t \approx \frac{\overline{X}_1 - \overline{X}_2}{\sqrt{\frac{s_1}{n_1} + \frac{s_2}{n_2}}}$$

z equals score minus the mean divided by the standard deviation t equals score (mean 1 minus mean 2 gives us the difference which we consider to be a score) divided by the standard deviation (the standard deviation of both groups; remember that the square root of variance is the standard deviation).

You see then that the t is a z.

Why is the variance of each group divided by the number of subjects in the group?

Remember, Gosset's goal was to make it difficult to researchers to find significance if their sample size was small.

Next we will try to understand the logic of the t-curve, the curve of no difference.

Drama
Me minus me equals 1

Professor Lilly Prydum, a statistician, decided to run a simple experiment to test a model for guessing if two means came from one specific group, or they came from two different groups.

Quite convoluted, you say.

She invited Memy Tallibum, Ph.D. in Education, to be part of the experiment and try to guess if the difference of two means came from the same group, or not. The subjects were 30 students from Trenton College. They were asked to take a test of 120 questions on Chinese geography, mythology, and culture. The scoring of the test was done by a computer, and the recording of the data was done in such a way that the subjects remained anonymous. The experiment lasted 35 days. Subjects were asked not to read about China during this time period. They were also asked not to chat amongst themselves and not to compare notes.

Day 1 of the experiment. Time 9:00 in the morning. The test is given. Time allotted 1 hour. 10:00 a.m. the testing period is over. Students are asked to take a 30 minute break. During that time the exam papers are graded, and the score is recorded next to each subject. Dr. Tallibum gets a copy of the results and studies them with care.

At 10:30 the students are called back in again, and are given the same test, the one they took 30 minutes earlier. At 11:30 the testing session is over. The subjects are thanked, and asked to be back the ext day. "Please be back here at 9:00 in the morning sharp. And remember, no reading on China".

The students leave, their exam papers are graded, and the score of each student is recorded. Each subject now has two scores. The score from session 1 and the score from session 2. Dr. Tallibum gets a copy and scrutinizes it. As expected the difference between the first session and the second is very small, and for most students it is zero.

The next day all students are present, Dr. Tallibum is present, and the experiment proceeds as planned.

Session 1, the students are given the same test they took previous day. Session 2, the students are given the same test. Scores are recorded and are handed to Dr. Tallibum who watches like a hawk. At the end of the second day Dr. Tallibum is not surprised to see that, once again, the difference between the first and the second session is very small, close to zero. Students are reminded that the experiment will last 35 days, and asked to "come back tomorrow".

On day 16 of the experiment Dr. Tallibum is surprised to see that the difference between the first and the second session is substantial. Students did not score as well in the second session. The explanation came in the evening as she was watching the news. Pop star Mickey Wiggletuchs had died. Students must have heard about this during the break between the two sessions and were emotionally disturbed. That influenced their performance in the second session.

On day 33 Dr. Tallibum was shocked to see that the difference between session 1 and session 2 was so great. Scores in the second session were so low! Again the explanation came in the evening news. The stock market had crashed. Apparently the students heard of this disaster during the break between the two session. Most likely parents called and told their children that there would be no money to pay the college fees.

Day 34, the day before the last day of the experiment, rolls smoothly. At the end of the second session, Dr. Tallibum is asked to leave the exam room.

Students are given the same instructions as usual. They are told to be present as usual, for the last day of the experiment.

Dr. Tallibum is left in the dark, she does not know what the students were told.

Day 35, the last day of the experiment.

Session 1. Students are taking the test as in the past and Dr. Tallibum is present, as in the past. During the break the exam papers are graded and scores handed to Dr. Tallibum.

At this point, there is a dramatic change in the procedure. Dr. Tallibum is asked to leave the exam room. She is led to the elevator and taken to the biology laboratories on the top floor, where there are no windows. She is asked not to leave this lab, not to make any phone calls, and wait there for an hour. To make sure, she is guarded by the graduate students of the biology lab.

At 11:30 the second session of the experiment is over. The students are told that the experiment came to an end, that there was no need for them to come back again, and were given a brief talk as to the purpose of the experiment.

The papers were graded, the scores were recorded and Professor Lilly Prydum, the experimenter, took the elevator to the Biology Lab and handed the results to Dr. Tallibum.

Dr. Tallibum, she said. In the second session we prevented you from seeing who were the subjects. As you realize we may have used the same subjects, the students from Trenton College, or we may have used another group. We could have used workers from the Physical Plant of the College, students from Trenton High, or senior citizens from the local senior citizens club. Your task is to guess whether the group that took the test during today's second session of the experiment were the Trenton College students that you watched for 35 days, or another, different group.

Dr. Tallibum looked at the data sheet. It contained the difference between session 1 and session 2 for 35 days. The difference score of day 35 was the score in question. Instinctively, her eyes ran up and down the list of differences. She was trying to find if the score of day 35 has occurred in the past 34 days. Not only that. If it has occurred, how often has it occurred. If she finds that it has occurred many times, she will conclude that the students in the second session of today were the same students, that is the Trenton College students.

If the difference score of today is not on the list, if it has never occurred in the last 34 days, then the wisest guess would be that in the second session of today it was not the Trenton College students that took the test.

It will be more difficult to guess if the difference of day 35 has occurred a few times, very few times. That's where Dr. Tallibum will resort to her knowledge of statistics. Guess what. She will draw Goddess Normal Curve and pray. Let's listen to her reasoning out loud:

(continued) Dr. Tallibum's soliloquy

The solution to my problem must be in the normal curve, as always. What do I have here. I have a difference between two means but I do not know if the two means came from the same group, or from two different groups. I want to guess wisely. I will start by placing 0 in the middle of the normal curve. This is the curve of no difference. I assume that this curve graphs differences between means that come from a known case, a case in which all means come from the same group, the same people. If the difference that I am not sure about is 0 or close to 0, I can safely conclude that this difference comes from the same group of people, i.e. that in both sessions of day 35 it was the Trenton College students that took the test. If the difference in question is large ...ay, there's the rub. Whether it safer to say same or not say, and suffer the slings and arrows of rough ridicule, or to shut up and by going against my promise to participate, my participation end. But behold, there is light. I can compute the standard deviation of this no-difference curve and reason. I can engage in the archetypal dance.

Behold Dr. Tallibum dancing around the biology lab to the amazement of the bio graduate students.

I have a difference between the two means. I hold this difference up, wave it in the air, I baptize it score. Then I wear my glasses and stick my nose on the normal curve, running up and down the line with standard deviations on it, and mumble: where does this score fall? Where does this score fall?

I then use the z formula and find where exactly my score - oops, our difference - falls. If it falls within -1.96 and +1.96 then this score occurs frequently, 95% of the time, in this case of known no difference. If this score, difference, falls beyond 1.96 then it is a rare occurrence.

In this case the wisest guess is that it does not come from the same group, that is, in the second session it was not the Trenton students but another different group of people.

I hear a question. Speak up. *Can she find out what was this new* group *is*, you say. Yes, the Oracle of Delphi should be able to tell you that.

Drama
The long jump

The prototypic experimenter is standing in the middle of the normal curve holding his score or difference in his hand. He is focusing on 1.96 and is about to jump beyond it. If he succeeds in jumping over 1.96, he gets a medallion (his degree perhaps); if he does not succeed he lands on his behind, and gets a kick on the same.

Talking seriously now. All statistical tests are based on the same logic we developed above. So I advise you to read my theatrical masterpieces carefully, if you want to get the concepts, and be able to reason in statistics, and solve the problems with only 5 formulas.

What are these 5 formulas?

The five formulas we need to know:

$$s^2 \approx \frac{\Sigma(X-\overline{X})^2}{n}$$

$$z \approx \frac{X-\overline{X}}{s}$$

$$SEM \approx \frac{s}{\sqrt{n}}$$

$$t \approx \frac{\overline{X}_1 - \overline{X}_2}{\sqrt{\frac{s_1}{n_1} + \frac{s_2}{n_2}}}$$

$$F \sqrt{\frac{MS_{between}}{MS_{within}}}$$

An example of the t-test

A sports physiologist suspected that bouillon cubes may improve the performance of runners. She got this idea when she read a 1996 experiment of mine in the Journal of Physiology and Behavior. She randomly selected 12 U of I male undergraduates and randomly assigned them to either group 1 who were given a cup of soup with a bouillon cube, or group 2 who were given a cup of chamomile.

She subsequently asked the students to run a 100 meter course and recorded the time they took to reach the finish line. Here is the layout of the experiment.

GROUP 1 BOUILLON	GROUP 2 NO BOUILLON
SUBJECT 1	SUBJECT 7
SUBJECT 2	SUBJECT 8
SUBJECT 3	SUBJECT 9
SUBJECT 4	SUBJECT 10
SUBJECT 5	SUBJECT 11
SUBJECT 6	SUBJECT 12

Note that a subject belongs to only one group, so the groups are independent. If subject 1 is John, and he belongs to the first group, he does not participate in the second group.

When the data were collected, she analyzed them by using the t-test. We say *she ran a t-test*. The data and analysis are presented in the next table:

	Bouillon	No Bouillon
	34	48
	35	47
	39	45
	30	47
	39	46
Mean	36.17	46.33
Varianc	12.47	1.22
df	10	
t	6.14	
p	<0.05	

What do you mean by df?, you say. You look unhappy too.

In order to develop the concept of degrees of freedom, df, we will run a *tought experiment* (thought experiment) as Einstein used to say. The benefit will be that you will not need to memorize any of the many formulas for degrees of freedom. Ever.

Drama
Mean Prophet

I take a sheet of paper from my printer and cut it into four equal pieces. I take a pencil and write the number 3 on the first piece, I write number 4 on the second piece, number 8 on the third piece, and lastly number 5 on the fourth piece of paper. I place all four pieces of paper in a shoe box and put the cover on it so that the pieces of paper cannot be seen. On another sheet of paper I keep a record of these four numbers. Then I add them up 3+4+8+5=20. Then I calculate the average or mean.

20 divided by 4 equals 5. The mean is 5. I destroy this piece of paper.

I take another sheet of paper and I write on it the following:
Mean=5 n=4

I stick this paper on the shoe box for anyone to see.

I ask Jerry, the English major who is sitting in the lounge, to come into the room. I explain to him that there are 4 pieces of paper in the box, each having a number on it. The mean of those numbers is 5. That is all he is told.

I begin by asking Jerry:
Jerry, close your eyes, stick your hand in the shoe box and pick one of the four pieces of paper.
Jerry does that. Before he draws his hand out of the box I ask him to guess the number he has picked. Jerry giggles.

I am not a magician, he says.

Open your eyes and read out the number.

Eight, he says.

I lay the piece of paper with the number 8 down on the table so that it can be seen at glance at anytime.

Now Jerry, stick your hand again in the shoebox and pick another piece of paper. Jerry does so. Before he draws his hand out of the box, I ask him to guess the number he has picked.

Again Jerry giggles and mumbles: *No way!*

Open your eyes and read the number out loud.

Five, he says.

I lay the piece of paper on the table next the first one that Jerry picked, the one that has the number 8 on it. It can easily be seen. There are two pieces of papers on the table now with the numbers, 8 and 5.

I repeat the procedure for the third time.

Jerry, can you guess what is the number that you drew?

He smiles faintly and simply shrugs his shoulders.

Open your eyes and read the number out loud.

Four, he says.

Again, I lay the piece of paper with the number 4 on the table, next to the other two. There are three numbers on the table now: 8 and 5 and 4.

We are ready to repeat the procedure, I say. Close your eyes and stick ...

Before I can finish my sentence, Jerry says:

Number 3.

Great! Jerry knows simple arithmetic. He added up the numbers he had already drawn.

8+5+4=17. There is only one number which, if added to 17, will give 20. That is the number 3. There is only one number that divided by 4 will give us a mean of 5. That number is 20.

What am I to get out of this story? you say.

Without knowing the mean and the n (how many numbers) you could not guess any of the numbers. They were free to vary. Several arrays of four numbers could give us a sum of 20. For example 11+1+2+6=20, or 5+ 6,+8,+1=20. Because I calculated the mean and showed it to you, and I also told you how many numbers there are in the box, one of the numbers is not free to vary.

In statistical language we say: We lose one *degree of freedom* every time we calculate a mean.

> We lose one degree of freedom for every mean that we calculate.

Remember this.

The degrees of freedom in this case is 3, that is
4-1=3.

Formally we write it as follows:
df=3.

In another situation that we have two groups of 10 subjects each, 20 total, and we calculate two means the degrees of freedom is 18. That is, we subtract 1 for each mean we calculate. This is simple to remember. I am confident that you understand this concept at the gut level, not just repeating my words like a parrot.

As I promised you, we will push aside almost all the formulas that otherwise you would have to memorize.

It is logical, isn't it. If you know the concept and you know what you are doing, you do not need a formula to tell you what to do step by step.

Back to our discussion of the t-test.

Definition: *t obtained.*.

That is the result of solving for t. In other words when you run a t-test you find a t value. A third way of saying this is: the result of analyzing your data using the t formula.

Definition: *t required.*

The required t is the value contained in the t-table which is found in the end of every statistics book, including the one you are reading now (see Appendix 1).

Remember, the t-table lists the modified 1.96 that Gosset published. It lists the recalculated values for 1.96 depending on degrees of freedom.

To find the required t, you first calculate the degrees of freedom. You are an expert in calculating degrees of freedom (df). No formulas needed, not for us who learn statistics by acting in soap operas.

In the present example of the t-test (page 114) we have two groups of 6 subjects each, total of 12 subjects. Since in computing the t we need to first compute the mean of each group, two means, we lose 2 degrees of freedom. How many scores go into the calculation of the t? All of the scores. That is,12 scores, minus 2 equals 10. Therefore, df=10. Now we go to the t table in Appendix 1 and run our finger down the left column which is labeled df. We stop at 10. Then we draw out finger horizontally until we reach the column that is labeled 5% or 0.05. We copy the value we find at the tip of our finger. This is the required t.

We compare this with the obtained t, i.e. the one that we calculated. If the obtained t is larger than the required t, we have significance. We say that our finding (the difference between the two means) is reliable or significant. That means that we trust that, if we run the same experiment again, we will find a difference again.

We formally write this as follows:

The difference between the means of the two groups is significant ($p < 0.5$).

By that we mean that the finding we are reporting is reliable, but there is still a chance that it may not be "real". That chance is less than five per cent. Scientists around the world have agreed to accept findings for which the probability of being chance events and not "real" is less than five percent. You understand correctly, there is no absolute certainty in experimental natural science. Findings are taken to be "true" on a probability basis. You see that boring, compulsive statistics borders on philosophy if approached from the correct angle.

One last remark. It is really unwarranted to speak of truth in dealing with phenomena in the empirical, material world. We can only speak of truth in the formal, logical and mathematical sciences.

Two plus three equals five. This is true. Two plus three is six, is false. It makes no sense to say that the statement: "Valium at doses of 2, 5, 10, and 20 mg reduces anxiety", is true. It is simply reliable and there is a probability attached to it, no matter how small, that it may not be so.

Chapter 6

Analysis of variance

Analysis of variance
One-way ANOVA

Analysis of variance is used by scientists in order to analyze data from experiments of literally unlimited experimental designs. It is most popular and dominant statistical test in the biological and social sciences. The complexity of these designs ranges from very simple to frighteningly convoluted. The formulas are so many that no statistical book contains all of them.

As I promised you, we will navigate through this ocean with no formulas. We do not need them. If we understand what we are doing, if we get the concepts involved, we do not need formulas. Do you need a map in order to work around your kitchen?

Hurray, here we launch the big ocean liner, ANOVA!

What is ANOVA, what do we do in Analysis of Variance?

We analyze variance.

That is tautologous, you say.

Ok, we partition the variance. That is pretty much what we do.
Variance I know well, you say.
Analysis of Variance you know pretty well, I say.

Yes. Variance we know so well.

$$s^2 \approx \frac{\Sigma(X-\bar{X})^2}{n}$$

Remember? The sum of squared deviations of each score from the mean, and all of this divided by n, the number of scores that went into the calculation, i.e., here, the number of all scores.

Why do you say "here", you ask.

Good observation. n does not always represent the number of scores in an experiment. It is the number of observations, that is a safer way to say this. More of that soon.

I was saying that what we do in ANOVA is analyze the variance in our data, more specifically, we *partition* the variance.

What is the result of this analysis? Is it a t?

Something like a t. A modified t, I would say.

You said earlier that t is a modified z?

---
t is a modified z
F is a modified t ?

You are very observant. Yes. As I told you earlier, statistics is like a pyramid. Discoveries are based on earlier discoveries.
There is continuity. If you brush aside the many formulas and concentrate on concepts, you get a marvelous view of the edifice of statistics. Then you are in command. You can take decisions, be in a position to critically view experiments, defend yourself against criticism that is thrown at you, and ultimately add this knowledge to your personal philosophy.

Drama
Clip his tail

Fisher was determined to clip Gosset 's tail a bit.

That Gosset, *he thinks he is smart. He rides high in the world with his stupid t-distribution. Big deal! All he did was to add one puny column on the left of the normal distribution. The df column. Oh, yes, ok, ok... He did recalculate the z, big deal.*

Fisher had been scratching his head for months, engaging in obsessive dialogues with Gosset, downgrading his achievement but deep down he knew he was jealous.

I got to come up with something myself. What if I add another column to the normal distribution. Another column of what? Not df again? If not df, what then. I got to be more original. How about another line on top of the t table.

He did. The df Between.
Good Knighthood, Sir Fisher.

The result, the endpoint of ANOVA, is F. This is in honoring Fisher who developed ANOVA. We calculate the F by the so called F ratio which is:

$$F\sqrt{\frac{MS_{between}}{MS_{within}}}$$

We read this as follows:

F equals mean square between, divided by mean square within.

Remember that mean square is another way of saying variance. So you should not be worried with the F ratio.

What about between and within? you say.

That is easy. In computing the variance between we calculate a variance. We line up the means of the various groups in our experiment, we treat them like scores, and figure out the variance.

You are kidding, you say. *I have seen terrifying formulas for even the simplest ANOVA.*

You are correct.

Now the second part of your question. Mean square within. Easy again. You already know it. We compute the variance of the first group, and write it down, then we compute the variance of the second group and write it down, then we do the same for all the groups in our experiment. The number of groups can vary from 2 to as many as you wish. In the end we simply add these variances. That gives the variance within.

Amazing! you say. *No new formulas for ANOVA!*

You can use ANOVA right away. All you needed was to get the concepts of variance between and variance within.

What about partitioning the

variance? you say.

Variance between and variance within, if added together, give us the Total variance.

Total variance can be computed if you calculate the variance of all the scores of all your groups, disregarding what group a score came from. Again, all you need is the familiar variance formula.

That is unbelievable, you say. *I am confused. Every statistics book gives ANOVA summary tables.*

Yes indeed. That is the convention, but it is not necessary in order to compute the F ratio. In practically every case in which step by step instructions are given (without first developing the concepts) things acquire an aura of awesome complexity and difficulty, and, I am afraid, fake importance.

Because you and I cannot go against the whole world, let's take a quick look at the typical presentation of ANOVA. Up until the sixties, journals were including ANOVA summary tables in the publications. As I said, this gives a publication the semblance of quantitative science, but, alas, only a semblance.

ANOVA SUMMARY TABLE
One-way ANOVA

SOUR	SS	df	MS	F	p
Between					
Within					
Total					

We know every term on this table. Within is also called the *error* term. The error term is the term which goes in the denominator of the F ratio. In more complex designs, the error term may be other than the within term.

Review the five formula's that are needed for virtually all parametric statistics. Do not simply memorize them, look into them conceptually.

$$s^2 \approx \frac{\Sigma(X-\bar{X})^2}{n}$$

$$z \approx \frac{X-\bar{X}}{s}$$

$$SEM \approx \frac{s}{\sqrt{n}}$$

$$t \approx \frac{\bar{X}_1 - \bar{X}_2}{\sqrt{\frac{s_1}{n_1} + \frac{s_2}{n_2}}}$$

$$F\sqrt{\frac{MS_{between}}{MS_{within}}}$$

An example of ANOVA

A biologist wanted to see if quantity of vitamin C in diet may reduce body weight.

He randomly selected 15 male rats and randomly assigned them to the following three groups.

Group1 10 mg, Group2 20 mg, and Group3 30 mg.

He added this vitamin in the food of the rats daily for 30 days.

On the 30th day he weighed the rats. Here are the data and the ANOVA table.

GROUP 1 - 10 mg	GROUP 2 - 20 mg	GROUP 3 - 30 mg
200	204	214
203	210	220
199	214	225
190	219	220
204	211	229
Mean1 = 199.20	Mean2 = 211.60	Mean3 = 221.60
s = 5.54	s = 5.50	s = 5.68

ANOVA SUMMARY TABLE

Source	SS	df	NS	F	p
Between	1259	2	629.6	20.24	<0.0001
Within	373.2	12	31.10		
Total	1632	14			

The analysis showed that we have a significant effect. The differences between the means are significant, i.e. reliable. The p value is less than 1 in 10000. This means that the probability that the difference we report is a chance event (and not the result of our treatment of giving rats vitamin C) is less than 1 in 10000.

How did you compute the degrees of freedom, df, you ask.

You know this, if you know the concepts of Between, Within, and Total.

We said in order to compute the variance between we line up the means of all the groups and treat them as scores, and calculate the variance. How many means we have here? We have three means. We really consider them as scores here. In order to calculate the variance of three numbers, we must first calculate the mean. By calculating the mean we lose 1 degree of freedom for every mean, remember? So our df for the between term is 3-1=2.

We calculate variance within as we said above. We calculate the mean of the first group and then the variance of this group. Then we do the same for the second group, and then the third group. We add these 3 variances and this gives us the variance within. Since we compute 3 means in the process of calculating the variances, we lose 3 degrees of freedom. How many scores went into the calculations of variance within? All the sores, that is 15. Our degrees of freedom then is df=15-3=12.

Easy, no formulas needed, because we understand the concept of df and also variance within.

Lastly, we compute the df of Total as follows:

In order to compute the Total variance, we said, we take all of the scores of all the groups disregarding what group each score comes from. In order to calculate this variance we must first calculate the mean, therefore we lose 1 degree of freedom. How many scores went into the calculation of the Total variance? All of the scores, here 15. Our degrees of freedom for Total then is 15-1=14.

Note that adding up the df for Between and Within we find the df for Total. That is what we meant by partitioning. We said we partition variance. Here we partition the Total variance into variance between and variance within. Indeed, verify that SS between plus SS within equals SS Total. That is,
1259 + 373.2 = 1632.2

How did we get the p value? you ask.

I will be very practical here. As was the case with the t-distribution, here too, there is a table with the F values (see Appendix).

These are in a way the recalculated 1.96, that is the point on the curve beyond which 5% of the curve lies.

In the case of the t curve we entered the table with the degrees of freedom and found the required t at the 5% level. In the present case, we enter the F table with the degrees of freedom for Between (in the present experiment df=2) and also the degrees of freedom for Within (in the present experiment df=12). We locate the F on the table. This is the required F.

Then we compare the obtained F (the one we calculated, look at he summary table) to the required F. As in the case of the t-test, if our obtained F is greater than the required F, we have significance. We then say $p<0.05$.

I understand how we calculate df without formulas, however I do not see why we need it.

I see why you are confused. As I said this is what happens every time we try to teach in a mechanistic, compulsive, step by step way. The general practice of working with formulas blindly, and using the ANOVA summary table, often prevents the student from seeing what is going on.

As I said at the beginning, the ANOVA summary table is not needed. What you need to do is simply calculate the variance, and compare the variances, that is the F ratio. df is simply the n in the variance formula.

---

df is simply the n in the variance formula

I understand variance Between, variance Within, and df. However, I do not see why the F ratio can detect significance, you say.

A very important question, if indeed we are sincere when we say we want to understand the concepts and the logic of statistics.

Drama
Master of the waves

It is a beautiful, cool, calm day in Puerto Rico. You are sitting in a San Juan small café, nested on the rocks overlooking the magnificent Atlantic ocean. You are happy, sipping your coffee, Bacardi on the side, and slowly nibbling on a sinfully sweet piece of PR cake. It is quiet, the only thing you hear is the rhythmic sound of waves gently breaking on the foundations of the cafe. Suddenly you hear voices; it sounds like people are arguing. Soon their voices become loud enough, you can clearly hear what they are saying.

You don't believe me? Look again.
See? I caused that wave.

There is much laughter.

Buddy, you are nuts, that's what I say. The only waves you cause is in your brain. You go and see a shrink!

The argument grows in intensity and the guy with the claims to supernatural powers, keeps throwing small stones into the sea. You decide to join the noisy group and get the argument straight.

Guys, I have the answer to your argument. I will show you who is right. For now, let the sea rest and calm down, just in case this guy has disturbed it. Come sit and have a cup of coffee.

Ten minutes later, you take the lot to the edge of the rocks.

First we will measure the heights of the next 40 waves and record these data, you say.

When forty waves have been recorded, you turn to the guy with the supernatural claims and say:

Ok, this is your show now.

The guy, his confidence somewhat deflated, picks up a stone and hurls it into the sea. All eyes are fixed on the base of the rock, waiting for the next wave. The wave comes and is recorded.

The moment of truth, you say.

The height of the wave is read out loud. It is not taller than any of the 40 waves previously recorded. The miracle worker receives a truckload of cosmetic epithets and soon the café slips back into the beatific serenity. Sleep hovers over your eyelids. You dream of conquistadors and fierce Carib Indians, of rituals and dances that humans created in an attempt to understand their world.

Cute story, but I still do not understand why the F ratio can indeed measure that we have significance, you say.

The forty waves provided us with what I call the "endogenous" variance, baseline, the variation in the heights of the waves that is present when no obvious cause can be seen. The wave after the action of the ambitious miracle worker was the presumed effect of his manipulation (in statistics we call this *treatment)*. Comparing the baseline with the claimed effect of his manipulation can give us support, or lack thereof, for a connection between what he did and the result we observed.

If the result which he claims that he caused by his manipulation is bigger than the natural, (endogenous or spontaneous variance), then we may say that he caused the effect by his manipulation.

A brief parenthesis at this point to make sure we understand what we mean by saying ratio. Alas, mechanistic methods of teaching arithmetic without development of concepts, often prevent the pupil from understanding that in division, what we do is compare two numbers: the numerator to the denominator. If you have 20 dollars, and I have 5 dollars, in dividing 20 by 5, I compare 20 to 5. You have 4 times more money than I have.

Our treatment causes variance between to increase?, you say.

Yes, let's see it in an example. A pharmacologist is testing a new drug (tentatively named Coolx) that is suspected to lower body temperature. He randomly selects 10 male college students and randomly assigns them to two groups. Group 1 receives Coolx, Group 2 receives a placebo (an inert substance that has no effect on physiology).

Here is the layout of the experiment.

Group 1 - Coolx	Group 2 - placebo
Subject 1 Subject 2 Subject 3 Subject 4 Subject 5	Subject 6 Subject 7 Subject 8 Subject 9 Subject 10

Before the experiment proper, the pharmacologist records the temperature of the subjects, in order to have the baseline temperature, the temperature that is present without any manipulation on the part of the experimenter.

Here is the baseline temperature (Celsius)

Group 1 - Coolx		Group 2 - placebo	
Subject 1	36.9	Subject 6	37.0
Subject 2	37.0	Subject 7	36.6
Subject 3	36.7	Subject 8	36.3
Subject 4	36.4	Subject 9	36.8
Subject 5	36.7	Subject 10	36.9
Mean 1 = 36.740		Mean 2 = 36.720	
Variance 1 = 0.053		Variance 2 = 0.077	

We will now calculate variance between.

Remember, in order to calculate variance between we line up the means and treat them as scores. We then proceed and calculate the variance of these scores.

Here we have two means
36.740 36.720

The variance of these two scores is 0.053. This is variance between.

The next table shows temperature after the administration of drug Coolex to Group 1

Group 1 - Coolx	Group 2 - placebo
Subject 1 36.9	Subject 6 37.0
Subject 2 37.0	Subject 7 36.6
Subject 3 36.7	Subject 8 36.3
Subject 4 36.4	Subject 9 36.8
Subject 5 36.7	Subject 10 36.9
Mean 1 = 36.740	Mean 2 = 36.720
Variance 1 = 0.053	Variance 2 = 0.077

Note that the mean in Group 1 decreased. It was 36.740 before giving the drug, it is 35.94 now. Also note that the variance of this group did not change.

Now the big moment has arrived.

Has the variance between changed? If yes, we will be convinced that variance between is sensitive to our manipulation, i.e. that it senses the effect of the drug.

As usual, in order to calculate variance between, we line up the means and treat them as scores. We then calculate the variance of these scores.

The means here are:

35.94 36.720

The variance is 0.3042.
This is variance between.

Let's compare this to variance between before our manipulation of giving the drug:

We saw above that variance was 0.053 .

Voila! After giving the drug, that is after our treatment, variance between changed. Variance within did not change.

Conclusion: The F ratio is sensitive to our treatment. It does so, because variance Between changes because of our manipulation, while variance Within does not change.

Why? you say.

Remember that variance measures the distance of scores from the mean. The mean can increase or decease but the distance of scores from the mean does not change. The scores move up or down *with* the mean.

Analysis of Variance

Factorial Designs
Two-Way ANOVA

The ANOVA that we discussed so far is called 'One-way ANOVA' or 'Single-factor ANOVA'.

Now we will consider two-way ANOVA or two-factor ANOVA.

The concepts we developed so far also apply to two-way ANOVA.

What do you mean by one-way, single-factor, two-way, or two-factor? you say.

Drama
Beam storm

Rutgers College. May 9, 1999, 9:00 in the morning. Two sections of Statistics 101 are in class; two adjacent classrooms, C120 and C121. USS Spaceship Enterprise flew over the two classrooms and locked on the bio readings of the students. Then, classroom C120 was bombarded with a X-Z-LOBX beam for 10 milliseconds. The security cameras recorded an almost imperceptible tilt of the head to the left, while the professor of Statistics, without being aware, wrote the same complex formula for MS 5 times. Two nanoseconds after classroom C120 was bathed in the benevolent X-Z-LOBX beam, classroom C121 was bombarded by the same X-Z-LOBX beam for 100 milliseconds. All students raised the index finger of their right hand and stuck it in their left nostril. The professor started reciting the t-table but stopped short in a deluge of laughter from the students.
The duration of the students' responses was recorded by the spaceship and instantly transmitted to Houston where a robot was waiting to manually enter the data on the layout of the experiment. The layout of the experiment was made public, the data not.

The layout of the USS Enterprise experiment

10 ms	100 ms
Classroom C120 Student 1 Student 2 etc	**Classroom C121** Student 50 Student 51 etc

This experiment is a one-way ANOVA design.

Why? Because each student was bombarded with one beam.
We also say that this design is a single-factor ANOVA.

Why?

Because each student was bombarded with a single beam. Another way of saying this is, that each score in this experiment is the result of one beam, one factor, or one treatment. You may also come across the term *one-way classification.*

> What we do to the subjects is called:
> *independents variable*
> or
> *treatment*

Now it will be easy for us to understand two-way ANOVA.

An example of a two-way ANOVA

A psychiatrist wanted to see whether a combination of wine and vitamin C may have an effect on depression.

He randomly selected 10 male patients, and also 10 female patients, and randomly assigned them in two groups: wine group, or vitamin C group.

The layout of this experiment is presented in the next table:

	wine	vitamin C
male	subject 1 subject 2 subject 3 subject 4 subject 5	subject 6 subject 7 subject 8 subject 9 subject 10
female	subject 11 subject 12 subject 13 subject 14 subject 15	subject 16 subject 17 subject 18 subject 19 subject 20

Look at subject 1. This subject is influenced by two variables. Male gender, and also wine. The score of depression that he will give, will be the result of these two factors. For this reason we call this type of experiment a two-factor experiment. The same, of course, holds for all subjects. They are, in a way, under cross fire. Two factors hit them.

The layout above can also be given in a more abstract form.

Variable A is gender, variable B is nutrition. Each variable has two levels, a1 a2 and b1 b2

	b1	b2
a1	subject 1 subject 2 subject 3 subject 4 subject 5	subject 6 subject 7 subject 8 subject 9 subject 10
a2	subject 11 subject 12 subject 13 subject 14 subject 15	subject 16 subject 17 subject 18 subject 19 subject 20

We say: We have two variables, A and B. A is gender, B is nutrition. Each of these two variables has two levels. a1, a2, and b1, b2 . Because in this experiment we use 2 variables with 2 levels each, we call this experiment *2 x 2 factorial*. We read this as follows:

two by two factorial.

The ANOVA summary table for two-factor experiments is the following:

ANOVA SUMMARY TABLE
Two-Way, 2x2 Factorial

Source	SS	df	MS	F	p
Between A					
Between B					
A x B *					
Within					
Total					

* Also called interaction

Things are getting complicated, I hear you say.

I say: You already know everything in this new ANOVA.

Our approach of understanding the concepts and not memorizing formulas has paid out.

Why do we have two Between terms, A and B?, you say.

Because here we have two variables: gender, and nutrition, i.e. A and B. We want to know if gender (being male or female) has an effect, and also if nutrition (wine or vitamin C) has an effect. Remember, the Between term is the term that senses the effects of our treatments.

The Within term we also know. It is the variance of each group separately. The sum of these variances.

The Total term we also know. It is simply the variance of all scores without regard to what group they came from.

The interaction term is a Between term for cells taken diagonally: mean for a1b1+a2b2 and mean a1b2+a2b1. Look at the layout to visualize this.

	b1	b2
a1	subject 1 subject 2 subject 3 subject 4 subject 5	subject 6 subject 7 subject 8 subject 9 subject 10
a2	subject 11 subject 12 subject 13 subject 14 subject 15	subject 16 subject 17 subject 18 subject 19 subject 20

What is new here is the concept of the interaction term. We need to develop this concept, so we get a gut feeling for it.

When you give two treatments to subjects, one of the things you want to see is whether the two variables interact with each other.

To begin developing the concept of interaction, let us consider a simple experiment:

We give 5 mg of an anti anxiety drug, such as diazepam, and find that this results in an increase in the time patients sleep. This increase is 2 hours.

Using different subjects, we find that 200 ml of wine increase sleep time by 1 hour.

Now if we give both 5 mg of valium and 200 ml of wine, is it sure that we will get 3 hours increase in sleep time? Perhaps yes, perhaps no. We know that drugs may interact and produce dramatic results, if given together. You may have heard of cases in which diazepam taken together with alcohol caused coma, and even death, because of *potentiation*.

Students find the concept of interaction difficult. For this reason I will give an example later.

For the purposes of calculation of this term in the ANOVA, there is no problem. The df, as you would expect, is the df of A x the df for B.

The SS you can calculate by subtraction. SS total-(SS Between A+SS Between B+SS within).

Alternatively, you can compute the SS for AxB the same way you calculated the between term, but here calculate two means diagonally; i.e.
mean for $a_1b_1 + a_2b_2$
and mean for $a_2b_1 + a_1b_2$.

Then we proceed with the calculation of the variance of these means.

The type of ANOVA design we are discussing here is called factorial, because in designing the experiment we produce all possible combinations.

In the above example we have:

Male - Wine, Male Vitamin C
Female - Wine, Female Vitamin C

Read this several times, it sounds like a nursery rhyme. There is a symmetry in it.

Visualizing the layout of factorial designs

You will often come across experiments that use these designs, and if you go to graduate school there is good chance you will use them in your research.

We need to be able to visualize the designs in order to understand them and evaluate them. A key part of the task of a scientist is to be able to critically evaluate the research of others. Regrettably, even reputable journals publish research that is not sound.

We have considered so far a 2x2 design. How do we visualize this? We see two characters (forget that it is the number 2 here) separated by the symbol x which stands for *times*.

We have two things, two variables, we therefore write down A and also B.

A   B

Now we look again at 2x2 and this time pay attention to what number we have. Here we have 2.
We therefore write
A
a1 a2

Then we look at the number after the x . It is also 2 (mind you it does not have to always be 2, it can be, 4, 10 any number).

We therefore write
B
b1 b2

To sum up:
a1b1 a1b2
a2b1 a2b2

Read this several times, it sounds like a nursery rhyme. There is a symmetry in it.

This is how we visualize a 2x2 factorial:

a1b1   a1b2
a2b1   a2b2

Now let us consider this: 2x3

How do we visualize this? We see two characters (forget that it is the numbers 2 and 3 here) separated by the symbol x which stands for *times*. We have two things, two variables, we therefore write down A and also B.

A   B

Now we look again at 2x3 and this time pay attention to what numbers we have. Before x we have 2. We therefore write:
A
a1 a2

Then we look at the number after the x. It is 3 (mind you, it does not have to always be 3, it can be 4, 10, any number).

We therefore write:
B
b1 b2 b3

This is how we visualize a 2x3 factorial:

a1b1   a1b2   a1b3
a2b1   a2b2   a2b3

Now let us consider this: 2x3x5

How do we visualize this? We see three characters (forget that it is the numbers 2 and 3 and 5 here) separated by the symbol x which stands for *times*. We have three things, three variables, we therefore write down A, B, and also C.

A   B   C

Now we look again at 2x3x5 and this time pay attention to what numbers we have. Before x we have 2 .
We therefore write
A
a1 a2

Then we look at the number after the x . It is 3 (mind you, it does not have to always be 3, it can be, 4, 10, any number).
We therefore write
B
b1 b2 b3

Then we look at the third number after the x . It is 5 (mind you it does not have to always be 5, it can be, 6, 28, any number).
We therefore write
C
c1 c2 c3 c4 c5

Examples of factorial experiments
Example 1 of a 2x2 factorial.

A pharmacology graduate student working on his thesis wanted to find whether a new chemical, srt-X, which has been shown to block serotonin, may be beneficial to schizophrenic patients. He was also interested to see if electroshock has an effect on these patients when combined with srt-X.

He randomly selected 20 schizophrenic patients, and randomly assigned them to 4 groups:

electroshock - srt-X,
electroshock-no srt-X
no electroshock - srt-X,
no electroshock-no srt-X

The layout of this experiment is:

	srt-X	no srt-X
electroshock	subject 1 subject 2 subject 3 subject 4 subject 5	subject 6 subject 7 subject 8 subject 9 subject 10
no electroshock	subject 11 subject 12 subject 13 subject 14 subject 15	subject 16 subject 17 subject 18 subject 19 subject 20

The layout in abstract form is:

	b1	b2
a1	subject 1 subject 2 subject 3 subject 4 subject 5	subject 6 subject 7 subject 8 subject 9 subject 10
a2	subject 11 subject 12 subject 13 subject 14 subject 15	subject 16 subject 17 subject 18 subject 19 subject 20

Variable A has two levels, a1 and a2, and variable B has two levels, b1 and b2.

The next table shows the data he recorded in running the experiment. The numbers represent scores on a psychiatric test measuring intensity of schizophrenic behavior. The higher the number the worse the condition of the patient.

THE SEROTONIN BLOCKER PLUS SHOCK EXPERIMENT

	b1	b2
a1	subject1 11 subject2 11 subject3 13 subject4 12 subject5 10 mean = 11.4 variance = 1.3	subject6 17 subject7 18 subject8 17 subject9 16 subject10 17 mean = 17 variance = 0.5
a2	subject11 15 subject12 14 subject13 14 subject14 16 subject15 15 mean = 14.8 variance = 0.7	subject16 20 subject17 19 subject18 18 subject19 20 subject20 18 mean = 16.9 variance = 1

ANOVA SUMMARY TABLE OF THE SEROTONIN BLOCKER PLUS SHOCK EXPERIMENT

Source	SS	df	MS	F	p
Between A	36.45	1	36.45	41.66	<.0001
Between B	120.05	1	120.05	137.2	<.0001
A x B *	2.45	1	2.45	2.8	>.05
Within	14	16	0.88		
Total	172.95	19			

* Interaction

I will first discuss the table in terms of the calculations we did.

First and most important, the degrees of freedom.

If you tell me the degrees of freedom in any ANOVA experiment, but without the use of formulas (I do also mean resorting to memory for the recollection of formulas - ban formulas!), I know you know what you are talking about. Calculation of the F is easy, high school arithmetic.

> If you tell me the degrees of freedom, I know you know how to analyze that data.

Why df for Between A is 1?

Because in order to calculate variance Between we line up the means, consider them scores, and calculate the variance using the one and only formula for variance (all the other formulas for variance that you may see around are derived from this formula. Statisticians get their kicks by producing equivalent formulas, of considerable complexity and ornamental value!). Now you and I know that in order to calculate variance, we must first calculate the mean. Every time we calculate the mean, we lose 1 degree of freedom. Because, in the present example we have 2 scores (never mind that they are means), we are left with 1 df. That is 2-1=1.

I do not understand why you say we have 2 means for A, you ask.

Good question. A has two levels here, a1 and a2. That is shock and no shock. You see, when we deal with variable A, we ignore variable B. In other words we reduce this part of the analysis to a one-way, singe-factor ANOVA.

Why df for Between B is 1?, you say.

For the same reasons as in the previous paragraph, B has two levels, b1 and b2, drug and no drug. There are two means (scores). In order to calculate the variance of these two scores, we must first compute the mean. We

therefore lose 1 df. So the df for B is 2-1=1.
Why df for AxB interaction is 1?

This is easy. Since df for A is 1, and df for B is also 1, the df for AxB is 1x1=1.

Why the df for within is 16?

This is simple, too. We said variance within is variance for the first group plus variance for the second group, plus variance for the third groups and so on. We have four groups here. In order to calculate the variance of each group we must first calculate a mean. The consequence of this is that we lose 1 df for every mean we calculate. How many scores go into the calculation of variance for group 1? Five scores. Therefore df for the first group is 5-1=4. We calculate the variance of the remaining 3 groups in a similar way. Since we have 4 groups here, the df for Within is 4x4=16.

Note: Checksum. The sum of df for A, B, AxB, Within, equals df Total

SS for A, B, AxB, and Within equals SS Total.

Remember, we said that in ANOVA we partition variance.

Discussion of the experiment with the schizophrenic patients.

Look at the ANOVA Table again:

ANOVA SUMMARY TABLE OF THE SEROTONIN BLOCKER PLUS SHOCK EXPERIMENT

Source	SS	df	MS	F	p
Between A	36.45	1	36.45	41.66	<.0001
Between B	120.05	1	120.05	137.2	<.0001
A x B *	2.45	1	2.45	2.8	>.05
Within	14	16	0.88		
Total	172.95	19			

* Interaction

The p value (the probability that the difference or effect we are reporting may not be reliable or significant) for A is less than 1 in ten thousand (p<.0001|).

Variable A is electroshock in this experiment. This means that the two conditions, electroshock and no electroshock (condition 1: electroshock-drug, electroshock-no drug; condition 2: no-electroshock-drug, no- electroshock-no drug) produced a result, a significant difference. In other words those patients who received electroshock ended up different from those patients that did not receive electroshock.

The p value (the probability that the difference or effect we are reporting may not be reliable or significant) for B is less than 1 in ten thousand (p<.0001|).

Variable B is drug in this experiment. This means that the two conditions, drug and no drug (condition 1:drug-electroshock, drug-no electroshock, condition 2: no drug-electroshock, no drug-no electroshock) produced a result, a significant difference. In other words those patients who received the drug were different from those patients that did not receive
the drug.

The p value of AxB, the interaction is p>.05, We read this as follows: p greater than five per cent. This means that if we were to say that there was significant interaction between electroshock and drug, we would be running the chance of reporting an effect that is not reliable, not significant, meaning that if we or someone else were to do the same experiment again, most likely would not find a difference as we did.

As we said earlier the concept of interaction is a new one for us, and we need to understand it our way, at the gut level, as we are used to.

We will now consider an experiment in which the interaction is significant.

Example 2 of a 2x2 factorial experiment

A pharmacology graduate student working on his thesis wanted to find whether a new chemical, DOP-Y, which has been shown to elevate dopamine levels in the brain, may be beneficial to depressive patients. He was also interested to see if electroshock has an effect on these patients when combined with DOP-Y. He randomly selected 20 depressive patients, and randomly assigned them to 4 groups:

electoshock - DOP-Y,
electroshock-no DOP-Y
no-electroshock - DOP-Y,
no-electroshock-no DOP-Y

The layout of the pharmacology experiment

	DOP-Y	no DOP-Y
electroshock	subject 1 subject 2 subject 3 subject 4 subject 5	subject 6 subject 7 subject 8 subject 9 subject 10
no electroshock	subject 11 subject 12 subject 13 subject 14 subject 15	subject 16 subject 17 subject 18 subject 19 subject 20

The data he recorded are given in the next table.

High numbers indicate improvement.

The data of the pharmacology experiment

	DOP-Y	no DOP-Y
electoshock	52 49 55 50 54 mean = 52 variance = 6.5	33 35 30 32 34 mean = 32.8 variance = 3.7
no electroshock	40 39 38 41 39 mean = 39.4 variance = 1.3	10 13 17 11 14 mean = 13 variance = 7.5

THE PHARMACOLOGY EXPERIMENT
ANOVA SUMMARY TABLE

Source	SS	df	MS	F	p
Between A	1312.2	1	1312.2	276.25	<.0001
Between B	2599.2	1	2599.2	547.2	<.0001
A x B	64.8	1	64.8	13.64	<0.005
Within	76	16	4.75		
Total	4052.2	19			

In this table we see that A, B, and also AxB are significant.

Significance in A means that electroshock benefited the depressive patients.

Significance in B means that drug benefited the depressive patients.

Significance in AxB means that there was an interaction between electroshock and drug.

Not clear, you say,

You are correct.

Let us look at the graph of the interaction.

First we observe that the two lines, shock and no shock, are not parallel. Every time we have an interaction, the two lines are not parallel.

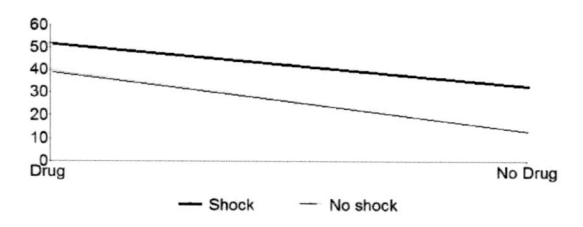

Every time we have an interaction, the two lines are not parallel.

How about getting to understand interaction at the gut level, not just with words? you say.

Let's do it. Look at the graph above (previous page).

First we will visualize the graph without the effects of the drug. In that graph the two lines would be parallel.

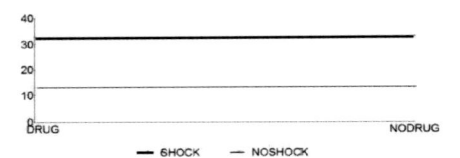

Now visualize the effect of drug as
a force pushing the lines up.

Logically we would expect to see
both lines pushed up while
maintaining the distance between
them, i.e. the two lines may move
higher on the graph, but they
should remain parallel.

However, in the present experiment
we saw that the drug has pushed the
no-electroshock line
disproportionately higher.

This is the concept of interaction.

Looking farther out in the ANOVA ocean

There is no limit in the number and complexity of experimental designs. ANOVA lends itself to analysis of these experiments, provided certain constraints are observed during the planning of the experiment.

The concepts in these designs are the same concepts that we developed in this book.

Complex designs are used more often in psychological and social science, and less often in biological, medical sciences. Quite frequently a t-test is used, often illegitimately. We should remember what we said in the beginning of this book. Statistics is a game for the statistician.

We use it only if we need it,
and should not allow our research
to be subjugated to statistics. This is
a vastly important issue. However, it
does not belong to a book like the
present one, whose only ambition
is to demystify statistics and allow
everyone to understand it.

Chapter 7

Repeated measures, correlated groups

We have already developed the concept of independence. In those experiments in which each subject is used only in one group or condition, we say that the groups are independent. So far in this book we have considered only independent-groups statistical designs and experiments.

In designs in which the groups are not independent, a subject is used in more than one group. That is, each subject experiences more that one treatment.

For example, John may first be given behavioral therapy, and later, several months later, he may also be given psychoanalytic therapy. The effects of the two therapies are then compared.

A variation of this arrangement is to match each subject with another subject on the basis of similarity in some measure. This is done to eliminate carryover effects that may, obviously, be present in giving one subject both treatments.

There are obviously advantages and disadvantages in choosing matched groups designs over independent groups designs. However this issue is beyond the goals of the present book. In general, independent groups designs are safer, and should, in my opinion, be preferred.

The concepts in matched groups designs are the same as those in independent groups designs. We will, therefore, confine ourselves to giving examples of these designs. First an example for t-test, and then an example for ANOVA repeated measures.

An example of t-test for matched groups

In comparing two new anti anxiety drugs, a pharmaceutical company selected 5 pairs of patients, each pair matched on the basis of their anxiety score.

Here is the layout and data of the experiment.

Subject	Difference Score
Tom and Susan	-1
Mike and Jim	-2
Anna and Mary	2
Jerry and Nick	1
George and Kim	-2

Mean for difference=0.4

The formula for the t-test for dependent groups is

$$t_{paired} \approx \frac{\overline{X}_D}{s/\sqrt{n}}$$

We read it as follows:

t for paired observations equals mean of differences divided by the standard deviation over the square root of the n. (The standard deviation divided by the square root of the n is the standard error of the mean, SEM, remember?)

Review

I juxtapose the three relevant formulas here
for you to compare:

$$z \approx \frac{X - \overline{X}}{s}$$

$$SEM \approx \frac{s}{\sqrt{n}}$$

$$t_{paired} \approx \frac{\overline{X}_D}{s/\sqrt{n}}$$

You know all of the terms of the
t-formula. You also recognize that it
is the same old story, our old
friend, the z formula.

t=+0.49 df=4

Entering the t-table with df 10 we
find that the required t=2.132

Our obtained t 0.49 is smaller
than the required, therefore we do
not have significance. We say that
the difference we observed is not
significant ($p>0.05$).

Example of ANOVA Repeated Measures

Four patients with damage in the hippocampus were treated with two new drugs in order to see if their memory improved.

Here is the layout as well as the scores of the experiment. High scores indicate improvement in memory.

Subject	Drug 1	Drug 2
Jim	5	8
Leo	3	9
Alice	6	7
Michel	4	6

ANOVA SUMMARY TABLE
Repeated Measures

Source	SS	df	MS	F	p
Between Columns	18	1	18	7.71	>0.05
Between Rows	3	3			
Error	7	3	2.33		
Total	28	7			

Entering the F table in Appendix 1 with df 1 and 3, we find an F of 10.12. This is the required F in order to have significance. Our obtained F (see ANOVA summary table above) is 7.71. It is less than the required F, therefore, we do not have significance. We say:

There was no significant difference between the means of the two conditions ($p>0.05$).

I see there are questions.

What is Between Columns?, you ask.

It is the usual Between variance that you know. The variance that our treatments produce. The variance of the means.

What is Between Rows?, you ask.

If you look at the layout above, you see that the rows are subjects, one subject per row. The mean of each subject is the mean of each row. The variance of these means are the variance between the rows.

Why you did not calculate an F for the Rows?, you ask.

There is no reason that I can think of, that would justify my wanting to know whether there is a statistical significant difference between subjects. That would be an absurd statement.

Once again you see that our conceptual approach allowed us to attack this design too, without the need for new formulas.

What is of course more important is the fact that we understand the logic of this design too.

We feel in command, comfortable to handle any questions on our statistics and research.

Complex designs

Elegant research avoids complex designs. However, you will not be spared of these monsters in your student or research life.

Let's get a whiff of these monsters. A psychiatrist wanted to see, if two new drugs improve the condition of depressive and schizophrenic patients.

He randomly assigned 4 depressive patients to Drug1 and Drug2 conditions. That is, each of the depressive patients will be serving as a subject in both the Drug conditions. This is a repeated measures design.

He did the same with the schizophrenic patients. He randomly assign 4 schizophrenic patients to Drug1 and Drug2 conditions. That is, each of the schizophrenic patients will be serving as a subject in both the Drug conditions. This is a repeated measures design.

As you see, here we have two independent groups (depressive patients, and schizophrenic patients) but each patient is given two treatments, that is he is tested repeatedly, i.e. in both drug conditions. We have a hybrid situation, you would say. Both independence and non-independence in the same experiment.

Here is the layout; X stands for scores.

	Drug1	Drug2
Depressive Patients		
Subject 1	X	X
Subject 2	X	X
Subject 3	X	X
Subject 4	X	X
Schizophrenic		
Subject 5	X	X
Subject 6	X	X
Subject 7	X	X
Subject 8	x	x

The analysis of data in complex designs like the above, is, as always, an operation involving the calculation of variance. The interpretation of the results of such an analysis is like the interpretations we considered in this book so far.

The calculations in such complex designs we leave up to computers and the compulsive.

Epilogue

I hope that you have enjoyed this
trip that we took together.

I am confident that what you
learned here will prove useful
in your courses, research, or
profession.

It would give me pleasure
to know that this book
added a tiny little *iota* to your
personal philosophy.

Those of you who are interested
in natural science, philosophy
and literature, may find my books
in these disciplines stimulating.

You can read my books in many
libraries in the USA (e.g. Harvard,
Columbia, Berkeley, Stanford)
and the UK (e.g. Oxford) as well as
in libraries of other countries.

APPENDICES

APPENDIX 1

Statistical Tables

The Normal Distribution Table

How to use this table,

The first left column begins with 00 and ends with **3.0** This column refers to standard deviation, or z.

The first, top line

0.00 0.01 0.02 0.03 0.04 0.05 0.06 0.07 0.08 0.09

simply adds decimals to the column. E.g. 1 on the left column increased to 1.01 1.02 all the way to 109 and then you shift to the next line, 1.1

Example: Place your finger at 1.9 on the first left column. Draw your finger horizontally to the seventh column. You read 0.4750 . This means that between the mean and standard deviation or z 1.96 47.5% of the area of the curve lies.

The Normal Curve Table

Z	0.00	0.01	0.02	0.03	0.04	0.05	0.06	0.07	0.08	0.09
0.0	0.0000	0.0040	0.0080	0.0120	0.0160	0.0199	0.0239	0.0279	0.0319	0.0359
0.1	0.0398	0.0438	0.0478	0.0517	0.0557	0.0596	0.0636	0.0675	0.0714	0.0753
0.2	0.0793	0.0832	0.0871	0.0910	0.0943	0.0987	0.1026	0.1064	0.1103	0.1141
0.3	0.1179	0.1217	0.1255	0.1293	0.1331	0.1368	0.1406	0.1443	0.1480	0.1517
0.4	0.1554	0.1591	0.1628	0.1664	0.1700	0.1736	0.1772	0.1808	0.1844	0.1879
0.5	0.1915	0.1950	0.1985	0.2019	0.2054	0.2088	0.2123	0.2157	0.2190	0.2224
0.6	0.2257	0.2291	0.2324	0.2357	0.2389	0.2422	0.2454	0.2486	0.2517	0.2549
0.7	0.2580	0.2611	0.2642	0.2673	0.2704	0.2734	0.2764	0.2794	0.2823	0.2852
0.8	0.2881	0.2910	0.2939	0.2967	0.2995	0.3023	0.3051	0.3078	0.3106	0.3133
0.9	0.3159	0.3186	0.3212	0.3238	0.3264	0.3289	0.3315	0.3340	0.3365	0.3389
1.0	0.3413	0.3438	0.3461	0.3485	0.3508	0.3531	0.3554	0.3577	0.3599	0.3621
1.1	0.3643	0.3665	0.3686	0.3708	0.3729	0.3749	0.3770	0.3790	0.3810	0.3830
1.2	0.3849	0.3869	0.3888	0.3907	0.3925	0.3944	0.3962	0.3980	0.3997	0.4015
1.3	0.4032	0.4049	0.4066	0.4082	0.4099	0.4115	0.4131	0.4147	0.4162	0.4177
1.4	0.4192	0.4207	0.4222	0.4236	0.4251	0.4265	0.4279	0.4292	0.4306	0.4319
1.5	0.4332	0.4345	0.4357	0.4370	0.4382	0.4394	0.4406	0.4418	0.4429	0.4441
1.6	0.4452	0.4463	0.4474	0.4484	0.4495	0.4505	0.4515	0.4525	0.4535	0.4545
1.7	0.4554	0.4564	0.4573	0.4582	0.4591	0.4599	0.4608	0.4616	0.4625	0.4633
1.8	0.4641	0.4649	0.4656	0.4664	0.4671	0.4678	0.4686	0.4693	0.4699	0.4706
1.9	0.4713	0.4719	0.4726	0.4732	0.4738	0.4744	0.4750	0.4756	0.4761	0.4767
2.0	0.4772	0.4778	0.4783	0.4788	0.4793	0.4798	0.4803	0.4808	0.4812	0.4817
2.1	0.4821	0.4826	0.4830	0.4834	0.4838	0.4842	0.4846	0.4850	0.4854	0.4857
2.2	0.4861	0.4864	0.4868	0.4871	0.4875	0.4878	0.4881	0.4884	0.4887	0.4890
2.3	0.4893	0.4896	0.4898	0.4901	0.4904	0.4906	0.4909	0.4911	0.4913	0.4916
2.4	0.4918	0.4920	0.4922	0.4925	0.4927	0.4929	0.4931	0.4932	0.4934	0.4936
2.5	0.4938	0.4940	0.4941	0.4943	0.4945	0.4946	0.4948	0.4949	0.4951	0.4952
2.6	0.4953	0.4955	0.4956	0.4957	0.4959	0.4960	0.4961	0.4962	0.4963	0.4964
2.7	0.4965	0.4966	0.4967	0.4968	0.4969	0.4970	0.4971	0.4972	0.4973	0.4974
2.8	0.4974	0.4975	0.4976	0.4977	0.4977	0.4978	0.4979	0.4979	0.4980	0.4981
2.9	0.4981	0.4982	0.4982	0.4983	0.4984	0.4984	0.4985	0.4985	0.4986	0.4986
3.0	0.4987	0.4987	0.4987	0.4988	0.4988	0.4989	0.4989	0.4989	0.4990	0.4990

The t-distribution Table
How to use this table.

The first column on the left of the table refers to degrees of freedom, df.

To find df simply subtract 2 from the number of scores you have in your experiment.

The first, top horizontal line refers to percentage of the t-distribution, the p or level of significance you wish to employ. The 5% level is adequate.

Example: If you have an experiment with 2 independent groups, 10 subject each, your df is 20-2=18

Suppose you analyzed your data and you found a t = 4.51

Now you enter the t-table. Place your finger on the first, left column on 18 (that is your df). Next draw your finger to the second column with the heading 0.05. You find t 1.734   Now compare your t value of t = 4.51 to the that of the table, which is 1.734 You see that your t is larger than the one in the table. This means that the finding of your experiment is significant, $p<0.05$

The t-distribution Table

Z	0.10	0.05	0.025	0.01	0.005	0.001
1.	3.078	6.314	12.70	31.82	63.65	318.3
2.	1.886	2.920	4.303	6.965	9.925	22.32
3.	1.638	2.353	3.182	4.541	5.841	10.21
4.	1.533	2.132	2.776	3.747	4.604	7.173
5.	1.476	2.015	2.571	3.365	4.032	5.893
6.	1.440	1.943	2.447	3.143	3.707	5.208
7.	1.415	1.895	2.365	2.998	3.499	4.782
8.	1.397	1.860	2.306	2.896	3.355	4.499
9.	1.383	1.833	2.262	2.821	3.250	4.296
10.	1.372	1.812	2.228	2.764	3.169	4.143
11.	1.363	1.796	2.201	2.718	3.106	4.024
12.	1.356	1.782	2.179	2.681	3.055	3.929
13.	1.350	1.771	2.160	2.650	3.012	3.852
14.	1.345	1.761	2.145	2.624	2.977	3.787
15.	1.341	1.753	2.131	2.602	2.947	3.733
16.	1.337	1.746	2.120	2.583	2.921	3.686
17.	1.333	1.740	2.110	2.567	2.898	3.646
18.	1.330	1.734	2.101	2.552	2.878	3.610
19.	1.328	1.729	2.093	2.539	2.861	3.579
20.	1.325	1.725	2.086	2.528	2.845	3.552
21.	1.323	1.721	2.080	2.518	2.831	3.527
22.	1.321	1.717	2.074	2.508	2.819	3.505
23.	1.319	1.714	2.069	2.500	2.807	3.485
24.	1.318	1.711	2.064	2.492	2.797	3.467
25.	1.316	1.708	2.060	2.485	2.787	3.450
26.	1.315	1.706	2.056	2.479	2.779	3.435
27.	1.314	1.703	2.052	2.473	2.771	3.421
28.	1.313	1.701	2.048	2.467	2.763	3.408
29.	1.311	1.699	2.045	2.462	2.756	3.396

30. 1.310 1.697 2.042 2.457 2.750 3.385
31. 1.309 1.696 2.040 2.453 2.744 3.375
32. 1.309 1.694 2.037 2.449 2.738 3.365
33. 1.308 1.692 2.035 2.445 2.733 3.356
34. 1.307 1.691 2.032 2.441 2.728 3.348
35. 1.306 1.690 2.030 2.438 2.724 3.340
36. 1.306 1.688 2.028 2.434 2.719 3.333
37. 1.305 1.687 2.026 2.431 2.715 3.326
38. 1.304 1.686 2.024 2.429 2.712 3.319
39. 1.304 1.685 2.023 2.426 2.708 3.313
40. 1.303 1.684 2.021 2.423 2.704 3.307
41. 1.303 1.683 2.020 2.421 2.701 3.301
42. 1.302 1.682 2.018 2.418 2.698 3.296
43. 1.302 1.681 2.017 2.416 2.695 3.291
44. 1.301 1.680 2.015 2.414 2.692 3.286
45. 1.301 1.679 2.014 2.412 2.690 3.281
46. 1.300 1.679 2.013 2.410 2.687 3.277
47. 1.300 1.678 2.012 2.408 2.685 3.273
48. 1.299 1.677 2.011 2.407 2.682 3.269
49. 1.299 1.677 2.010 2.405 2.680 3.265
50. 1.299 1.676 2.009 2.403 2.678 3.261
51. 1.298 1.675 2.008 2.402 2.676 3.258
52. 1.298 1.675 2.007 2.400 2.674 3.255
53. 1.298 1.674 2.006 2.399 2.672 3.251
54. 1.297 1.674 2.005 2.397 2.670 3.248
55. 1.297 1.673 2.004 2.396 2.668 3.245
56. 1.297 1.673 2.003 2.395 2.667 3.242
57. 1.297 1.672 2.002 2.394 2.665 3.239
58. 1.296 1.672 2.002 2.392 2.663 3.237
59. 1.296 1.671 2.001 2.391 2.662 3.234
60. 1.296 1.671 2.000 2.390 2.660 3.232
61. 1.296 1.670 2.000 2.389 2.659 3.229

62. 1.295 1.670 1.999 2.388 2.657 3.227
63. 1.295 1.669 1.998 2.387 2.656 3.225
64. 1.295 1.669 1.998 2.386 2.655 3.223
65. 1.295 1.669 1.997 2.385 2.654 3.220
61. 1.296 1.670 2.000 2.389 2.659 3.229
62. 1.295 1.670 1.999 2.388 2.657 3.227
63. 1.295 1.669 1.998 2.387 2.656 3.225
64. 1.295 1.669 1.998 2.386 2.655 3.223
65. 1.295 1.669 1.997 2.385 2.654 3.220
66. 1.295 1.668 1.997 2.384 2.652 3.218
67. 1.294 1.668 1.996 2.383 2.651 3.216
68. 1.294 1.668 1.995 2.382 2.650 3.214
69. 1.294 1.667 1.995 2.382 2.649 3.213
70. 1.294 1.667 1.994 2.381 2.648 3.211
71. 1.294 1.667 1.994 2.380 2.647 3.209
72. 1.293 1.666 1.993 2.379 2.646 3.207
73. 1.293 1.666 1.993 2.379 2.645 3.206
74. 1.293 1.666 1.993 2.378 2.644 3.204
75. 1.293 1.665 1.992 2.377 2.643 3.202
76. 1.293 1.665 1.992 2.376 2.642 3.201
77. 1.293 1.665 1.991 2.376 2.641 3.199
78. 1.292 1.665 1.991 2.375 2.640 3.198
79. 1.292 1.664 1.990 2.374 2.640 3.197
80. 1.292 1.664 1.990 2.374 2.639 3.195
81. 1.292 1.664 1.990 2.373 2.638 3.194
82. 1.292 1.664 1.989 2.373 2.637 3.193
83. 1.292 1.663 1.989 2.372 2.636 3.191
84. 1.292 1.663 1.989 2.372 2.636 3.190
85. 1.292 1.663 1.988 2.371 2.635 3.189
86. 1.291 1.663 1.988 2.370 2.634 3.188
87. 1.291 1.663 1.988 2.370 2.634 3.187
88. 1.291 1.662 1.987 2.369 2.633 3.185

89. 1.291 1.662 1.987 2.369 2.632 3.184
90. 1.291 1.662 1.987 2.368 2.632 3.183
91. 1.291 1.662 1.986 2.368 2.631 3.182
92. 1.291 1.662 1.986 2.368 2.630 3.181
93. 1.291 1.661 1.986 2.367 2.630 3.180
94. 1.291 1.661 1.986 2.367 2.629 3.179
95. 1.291 1.661 1.985 2.366 2.629 3.178
96. 1.290 1.661 1.985 2.366 2.628 3.177
97. 1.290 1.661 1.985 2.365 2.627 3.176
98. 1.290 1.661 1.984 2.365 2.627 3.175
99. 1.290 1.660 1.984 2.365 2.626 3.175
100. 1.290 1.660 1.984 2.364 2.626 3.174
∞   1.282 1.645 1.960 2.326 2.576 3.090

The F Distribution Table
How to use this table

The first, left column is the df of the denominator of the F ratio, or the error term.

The first, top line is the degrees of freedom of the numerator of the F ratio. Example:

In an experiment with 3 groups, of 10 subjects each, we have df between 2, and df within 27.

Suppose that we analyzed our data using ANOVA and we found an F=12.54

Now we enter the F table. Place your finger on the first left column at 27, then draw your finger to the third column which has the heading 2. You read the value 3.354

Now you compare your F which was F=12.54 to the F of this table which was 3.354 Your F is bigger, so the finding of your experiment is significant, $p<0.05$

The F distribution Table 5% significance level

	1	2	3	4	5	6	7	8	9	10
1	161.4	4199.5	215.7	224.5	230.1	233.9	236.7	238.8	240.5	241.8
2	18.51	19.00	19.16	19.24	19.29	19.33	19.35	19.37	19.38	19.39
3	10.12	9.552	9.277	9.117	9.013	8.941	8.887	8.845	8.812	8.786
4	7.709	6.944	6.591	6.388	6.256	6.163	6.094	6.041	5.999	5.964
5	6.608	5.786	5.409	5.192	5.050	4.950	4.876	4.818	4.772	4.735
6	5.987	5.143	4.757	4.534	4.387	4.284	4.207	4.147	4.099	4.060
7	5.591	4.737	4.347	4.120	3.972	3.866	3.787	3.726	3.677	3.637
8	5.318	4.459	4.066	3.838	3.687	3.581	3.500	3.438	3.388	3.347
9	5.117	4.256	3.863	3.633	3.482	3.374	3.293	3.230	3.179	3.137
10	4.965	4.103	3.708	3.478	3.326	3.217	3.135	3.072	3.020	2.978
11	4.844	3.982	3.587	3.357	3.204	3.095	3.012	2.948	2.896	2.854
12	4.747	3.885	3.490	3.259	3.106	2.996	2.913	2.849	2.796	2.753
13	4.667	3.806	3.411	3.179	3.025	2.915	2.832	2.767	2.714	2.671
14	4.600	3.739	3.344	3.112	2.958	2.848	2.764	2.699	2.646	2.602
15	4.543	3.682	3.287	3.056	2.901	2.790	2.707	2.641	2.588	2.544
16	4.494	3.634	3.239	3.007	2.852	2.741	2.657	2.591	2.538	2.494
17	4.451	3.592	3.197	2.965	2.810	2.699	2.614	2.548	2.494	2.450
18	4.414	3.555	3.160	2.928	2.773	2.661	2.577	2.510	2.456	2.412
19	4.381	3.522	3.127	2.895	2.740	2.628	2.544	2.477	2.423	2.378
20	4.351	3.493	3.098	2.866	2.711	2.599	2.514	2.447	2.393	2.348
21	4.325	3.467	3.072	2.840	2.685	2.573	2.488	2.420	2.366	2.321
22	4.301	3.443	3.049	2.817	2.661	2.549	2.464	2.397	2.342	2.297
23	4.279	3.422	3.028	2.796	2.640	2.528	2.442	2.375	2.320	2.275
24	4.260	3.403	3.009	2.776	2.621	2.508	2.423	2.355	2.300	2.255
25	4.242	3.385	2.991	2.759	2.603	2.490	2.405	2.337	2.282	2.236
26	4.225	3.369	2.975	2.743	2.587	2.474	2.388	2.321	2.265	2.220
27	4.210	3.354	2.960	2.728	2.572	2.459	2.373	2.305	2.250	2.204
28	4.196	3.340	2.947	2.714	2.558	2.445	2.359	2.291	2.236	2.190
29	4.183	3.328	2.934	2.701	2.545	2.432	2.346	2.278	2.223	2.177
30	4.171	3.316	2.922	2.690	2.534	2.421	2.334	2.266	2.211	2.165
31	4.160	3.305	2.911	2.679	2.523	2.409	2.323	2.255	2.199	2.153
32	4.149	3.295	2.901	2.668	2.512	2.399	2.313	2.244	2.189	2.142
33	4.139	3.285	2.892	2.659	2.503	2.389	2.303	2.235	2.179	2.133
34	4.130	3.276	2.883	2.650	2.494	2.380	2.294	2.225	2.170	2.123
35	4.121	3.267	2.874	2.641	2.485	2.372	2.285	2.217	2.161	2.114
36	4.113	3.259	2.866	2.634	2.477	2.364	2.277	2.209	2.153	2.106
37	4.105	3.252	2.859	2.626	2.470	2.356	2.270	2.201	2.145	2.0
38	4.098	3.245	2.852	2.619	2.463	2.349	2.262	2.194	2.138	2.091
39	4.091	3.238	2.845	2.612	2.456	2.342	2.255	2.187	2.131	2.084
40	4.085	3.232	2.839	2.606	2.449	2.336	2.249	2.180	2.124	2.077

The F distribution Table
41 4.079 3.226 2.833 2.600 2.443 2.330 2.243 2.174 2.118 2.071
42 4.073 3.220 2.827 2.594 2.438 2.324 2.237 2.168 2.112 2.065
43 4.067 3.214 2.822 2.589 2.432 2.318 2.232 2.163 2.106 2.059
44 4.062 3.209 2.816 2.584 2.427 2.313 2.226 2.157 2.101 2.054
45 4.057 3.204 2.812 2.579 2.422 2.308 2.221 2.152 2.096 2.049
46 4.052 3.200 2.807 2.574 2.417 2.304 2.216 2.147 2.091 2.044
47 4.047 3.195 2.802 2.570 2.413 2.299 2.212 2.143 2.086 2.039
48 4.043 3.191 2.798 2.565 2.409 2.295 2.207 2.138 2.082 2.035
49 4.038 3.187 2.794 2.561 2.404 2.290 2.203 2.134 2.077 2.030
50 4.034 3.183 2.790 2.557 2.400 2.286 2.199 2.130 2.073 2.026
51 4.030 3.179 2.786 2.553 2.397 2.283 2.195 2.126 2.069 2.022
52 4.027 3.175 2.783 2.550 2.393 2.279 2.192 2.122 2.066 2.018
53 4.023 3.172 2.779 2.546 2.389 2.275 2.188 2.119 2.062 2.015
54 4.020 3.168 2.776 2.543 2.386 2.272 2.185 2.115 2.059 2.011
55 4.016 3.165 2.773 2.540 2.383 2.269 2.181 2.112 2.055 2.008
56 4.013 3.162 2.769 2.537 2.380 2.266 2.178 2.109 2.052 2.005
57 4.010 3.159 2.766 2.534 2.377 2.263 2.175 2.106 2.049 2.001
58 4.007 3.156 2.764 2.531 2.374 2.260 2.172 2.103 2.046 1.998
59 4.004 3.153 2.761 2.528 2.371 2.257 2.169 2.100 2.043 1.995
60 4.001 3.150 2.758 2.525 2.368 2.254 2.167 2.097 2.040 1.993
61 3.998 3.148 2.755 2.523 2.366 2.251 2.164 2.094 2.037 1.990
62 3.996 3.145 2.753 2.520 2.363 2.249 2.161 2.092 2.035 1.987
63 3.993 3.143 2.751 2.518 2.361 2.246 2.159 2.089 2.032 1.985
64 3.991 3.140 2.748 2.515 2.358 2.244 2.156 2.087 2.030 1.982
65 3.989 3.138 2.746 2.513 2.356 2.242 2.154 2.084 2.027 1.980
66 3.986 3.136 2.744 2.511 2.354 2.239 2.152 2.082 2.025 1.977
67 3.984 3.134 2.742 2.509 2.352 2.237 2.150 2.080 2.023 1.975
68 3.982 3.132 2.740 2.507 2.350 2.235 2.148 2.078 2.021 1.973
69 3.980 3.130 2.737 2.505 2.348 2.233 2.145 2.076 2.019 1.971
70 3.978 3.128 2.736 2.503 2.346 2.231 2.143 2.074 2.017 1.969
71 3.976 3.126 2.734 2.501 2.344 2.229 2.142 2.072 2.015 1.967
72 3.974 3.124 2.732 2.499 2.342 2.227 2.140 2.070 2.013 1.965
73 3.972 3.122 2.730 2.497 2.340 2.226 2.138 2.068 2.011 1.963
74 3.970 3.120 2.728 2.495 2.338 2.224 2.136 2.066 2.009 1.961
75 3.968 3.119 2.727 2.494 2.337 2.222 2.134 2.064 2.007 1.959
76 3.967 3.117 2.725 2.492 2.335 2.220 2.133 2.063 2.006 1.958
77 3.965 3.115 2.723 2.490 2.333 2.219 2.131 2.061 2.004 1.956
78 3.963 3.114 2.722 2.489 2.332 2.217 2.129 2.059 2.002 1.954
79 3.962 3.112 2.720 2.487 2.330 2.216 2.128 2.058 2.001 1.953
80 3.960 3.111 2.719 2.486 2.329 2.214 2.126 2.056 1.999 1.951

The F distribution Table
81 3.959 3.109 2.717 2.484 2.327 2.213 2.125 2.055 1.998 1.950
82 3.957 3.108 2.716 2.483 2.326 2.211 2.123 2.053 1.996 1.948
83 3.956 3.107 2.715 2.482 2.324 2.210 2.122 2.052 1.995 1.947
84 3.955 3.105 2.713 2.480 2.323 2.209 2.121 2.051 1.993 1.945
85 3.953 3.104 2.712 2.479 2.322 2.207 2.119 2.049 1.992 1.944
86 3.952 3.103 2.711 2.478 2.321 2.206 2.118 2.048 1.991 1.943
87 3.951 3.101 2.709 2.476 2.319 2.205 2.117 2.047 1.989 1.941
88 3.949 3.100 2.708 2.475 2.318 2.203 2.115 2.045 1.988 1.940
89 3.948 3.099 2.707 2.474 2.317 2.202 2.114 2.044 1.987 1.939
90 3.947 3.098 2.706 2.473 2.316 2.201 2.113 2.043 1.986 1.938
91 3.946 3.097 2.705 2.472 2.315 2.200 2.112 2.042 1.984 1.936
92 3.945 3.095 2.704 2.471 2.313 2.199 2.111 2.041 1.983 1.935
93 3.943 3.094 2.703 2.470 2.312 2.198 2.110 2.040 1.982 1.934
94 3.942 3.093 2.701 2.469 2.311 2.197 2.109 2.038 1.981 1.933
95 3.941 3.092 2.700 2.467 2.310 2.196 2.108 2.037 1.980 1.932
96 3.940 3.091 2.699 2.466 2.309 2.195 2.106 2.036 1.979 1.931
97 3.939 3.090 2.698 2.465 2.308 2.194 2.105 2.035 1.978 1.930
98 3.938 3.089 2.697 2.465 2.307 2.193 2.104 2.034 1.977 1.929
99 3.937 3.088 2.696 2.464 2.306 2.192 2.103 2.033 1.976 1.928
100 3.936 3.087 2.696 2.463 2.305 2.191 2.103 2.032 1.975 1.927

APPENDIX 2

Frequently Asked Questions, FAQ

1. Q. I have an experiment with 2 independent groups. The scale of measurement is nominal. Can I use a t-test?
A. No, use non-parametric tests

2. Q. I have an experiment with 2 independent groups. The scale of measurement is interval. Can I use a t-test?
A. Yes

3. Q. I have an experiment with 2 independent groups. The scale of measurements is ratio. Can I use a t-test?
A. Yes

4. Q. I have an experiment with 2 independent groups. The scale of measurement interval. Can I use ANOVA?
A. Yes

5. Q. I have an experiment with 2 independent groups. The scale of measurement ratio. Can I use ANOVA?
A. Yes

6. Q. I have an experiment with 3 independent groups. The scale of measurement is interval. Can I use a t-test?
A. No. Use ANOVA

7. Q I have run an experiment with 4 independent groups. 20 male and 20 female subjects. The experiment aimed at showing that drinking red wine may lower cholesterol. The layout was as

follows:
Male-wine, male-no wine.
Female-wine, female-no wine
There were 10 subjects per group. I think this is a 2x2 factorial, but I do not understand why gender is variable. I did not manipulate it, obviously.
A. Gender is a special class of variable. We call this *subject variable*. While we do not manipulate it ourselves, some other agent has done it, in this case Nature, we analyze these experiments like regular designs. Only in the discussion we should be aware that subject variables are actually packages of variables, and we should be cautious in our conclusions.

8. Q. I have an experiment with 3 independent groups. My data are in the interval scale of measurement. I ran an ANOVA

and found significance. Does this mean that (a) mean1 is different from mean 2? (b) mean 1 is different from mean 3? (c) mean 2 is different from mean 3?
A. No. It means that your treatment has had an effect, but you do not know where the difference is. ANOVA does not tell you that. You need to run one of the so called *post hoc tests* or otherwise called *a posteriori tests,* in order to pinpoint where the difference is.

The most frequently used post hoc tests are: The *Newman-Keuls* test, *Tukey's test*, the *Scheffé* test.

9. Q. I have an experiment with 3 independent groups, ratio scale of measurement. Can I run post hoc tests without running ANOVA?
A. No. First run ANOVA. If you find significance, then proceed with post hoc tests.

10. Q. I have an experiment with 3 independent groups, ratio scale of measurement. Can I run t-test repeatedly so that I see which mean is significantly different from which mean?
A. No. Run a post hoc test.

APPENDIX 3

The equation which produces the normal distribution:

$$y = \frac{1}{\sqrt{2\pi}} e^{-x^2/2}$$

The five formulas that we need

$$s^2 \approx \frac{\sum \left(X - \bar{X}\right)^2}{n}$$

$$z \approx \frac{X - \bar{X}}{s}$$

$$SEM \approx \frac{s}{\sqrt{n}}$$

$$t \approx \frac{\bar{X}_1 - \bar{X}_2}{\sqrt{\frac{s_1^2}{n_1} + \frac{s_2^2}{n_2}}}$$

$$F \approx \frac{MS_{between}}{MS_{within}}$$

STORY TITLE	PAGE
My kids my fingers	13
No-number numbers	14
Who's bigger	17
Working magic with a Goddess	29
Mathematical sweat	38
Where does Basita fall?	53
An archetypal ceremony	59
Money in a Texas Hat	61
A Cap for Wisconsin Farmers	66
Apple-pie IQ	82
An archetypal ceremony II	86
Mercy Mr. Gosset	91
Me minus me equals 1	102
The long jump	112
Mean Prophet	117
Clip his tail	134
Master of the waves	148
Beam storm	160

Manufactured by
Amazon.ca
Bolton, ON